More Mathematical Quilts

No Sewing Required!

Diana Venters
Elaine Krajenke Ellison

Key Curriculum Press
Innovators in Mathematics Education

Project Editor	Joan Lewis
Editorial Assistants	Christopher David, Lori Dixon, Beth Masse, Jason Taylor
Editorial Consultant	Stacey Miceli
Mathematics Consultant	Larry Copes
Production Editor	Angela Chen
Copyeditor	Robert Fiske
Production Director	Diana Jean Parks
Production Coordinators	Mike Hurtik, Chris Pimentel
Original Series Designer (Text and Cover)	Kirk Mills
Cover Designer	Caroline Ayres
Composition, Technical Art	Interactive Composition Corporation
Printer, Prepress	Versa Press, Inc.
Executive Editor	Casey FitzSimons
Publisher	Steven Rasmussen

Key Curriculum Press
1150 65th Street
Emeryville, CA 94608
editorial@keypress.com
http://www.keypress.com

Printed in the United States of America 10 9 8 7 6 5 4 3 2 1 07 06 05 04 03 02 ISBN 1-55953-374-9

Contents

Note to Teachers

It is important for students to learn mathematics in a variety of ways, not only through problems in a textbook. As mathematics teachers who enjoy quilting, we saw that interesting quilts could be created to represent mathematics concepts. When we began to create those quilts, we realized that planning the design was also part of the learning experience. This book was created to help students share in that learning.

Mathematical Quilts, our first collection of quilts and classroom activities, was published in 1999. Now, more designs are available in this collection of *More Mathematical Quilts*. The quilts, and the activities based on them, allow students to learn mathematics in a unique way.

Learning Geometry and the van Hiele Levels

In their research on how students learn geometry, Dutch mathematicians Dina van Hiele-Geldof and Pierre van Hiele identified five levels of learning that students must progress through to succeed in geometry (*Begrip en Inzicht,* 1973). Students achieve the first level, visualization, when they recognize shapes by appearance. Students reach the second level, analysis, when they understand characteristics of a shape based on simple definitions and recognize shapes regardless of their orientation in the plane. Students attain the third level, informal deduction, when they are able to compare relationships among shapes and understand the minimal conditions needed to determine a shape. The activities in this book provide opportunities for students to work at the second and third levels of the van Hiele model. The fourth level, formal deduction, is needed for formal, proof-oriented geometry courses. The highest level, rigor, is not part of most high school curricula.

The van Hiele model is helpful in understanding the difficulty some capable students have in a formal geometry course. According to the van Hiele model, when students have not successfully progressed through the first three levels of geometry reasoning, they cannot be successful in formal geometry.

The activities in this book will help students improve their visualization skills and understanding of geometry relationships, and help them succeed at deductive reasoning in geometry.

How the Book Is Organized

The quilts in this book were all inspired by mathematical patterns. We hope that these quilts and the related activities will inspire you and your students to explore these patterns and enjoy the beauty we have discovered in the mathematics. The book is divided into four thematic sections, each highlighting a series of quilts with similar designs: "The Polygonal Fractal Quilts," "The Linear Fractal Quilts," "The Spatial Quilts," and "Even More Mathematical Quilts." After a short introduction to each theme, there is a description of each quilt and an explanation of how each quilt pattern was created. This is followed by instructions on how to actually make each quilt. The activities build on what students have learned from designing the quilts and extend the learning to related mathematics concepts. The activities include suggested research topics, technology activities for graphing calculators and computers, and Internet links through www.keymath.com/quilts. Teacher notes and solutions to activities conclude each thematic section. A bibliography of mathematics and quilting references appears in the Appendix, along with blackline masters of useful grids.

Suggestions for Classroom Use

You can use these activities to supplement or enhance topics in the geometry. The "Designing Your Quilt" sections encourage the use of color and make good mathematical art projects. They are suitable for enrichment, extension of classroom learning, or purely recreational mathematics. Use them before corresponding activities or independently. Although some of the activities within a section are sequential, most can be used separately. You will find information in the teacher notes sections relating the activities to particular quilt designs.

Students can work on the activities individually or in groups. You may choose to divide the activities in a section so that different groups work on different quilt designs and the mathematical activities that support these designs. Each group can then make a presentation to the class, explaining the mathematics and interpreting a specific design. You can assign students the research topics as individual projects, and they can present them to the class or at mathematics fairs or other school programs. Parents, administrators, community members, other teachers, and students will enjoy this unique method of presenting mathematics.

This is not a quilt-making book, though we have provided abbreviated instructions for sewing the quilts. For those students who wish to re-create the quilts without sewing, we suggest using colored construction paper or fabric. Students can trace and cut out the shapes, and glue the pieces onto poster board. Students with access to dynamic geometry software, such as The Geometer's Sketchpad®, or a computer drawing program may enjoy creating their designs on a computer. Don't be surprised, however, if interested students, or perhaps even parents, get excited enough to actually sew one of these quilts.

Diana Venters
Elaine Krajenke Ellison

The Polygonal Fractal Quilts

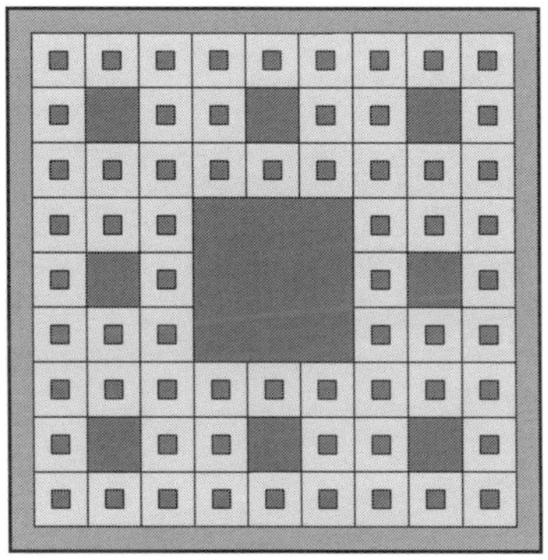

The Sierpiński Carpet Quilt

A Sierpiński Triangle Quilt

A FRACTAL IS THE RESULT OF A PROCESS THAT BEGINS with a geometric figure (called the *initiator*), applies a transformation to the figure to produce Level 1 (called the *generator*), transforms the generator in the same way to produce a new result, transforms that result in the same way, and so on for infinitely many steps.

Fractals are *self-similar*. That is, parts of the fractal look like the entire fractal, only at a smaller scale.

The quilts in this section show fractal designs completed to Level 3, 4, or 5. The two triangle quilts are based on a fractal known as the *Sierpiński triangle*. The first few levels of that fractal are shown on the next page.

The initiator of this fractal is a triangle. In the first iteration, the midpoints of the sides of the triangle are connected, creating four identical smaller triangles, and then the middle triangle is removed. This creates Level 1, a shape known as the *generator*. The process of creating this shape from the initiator is repeated to create, or *generate,* each level of the

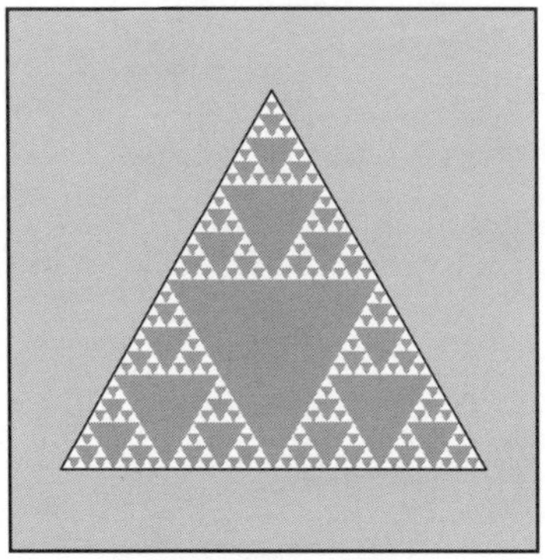

A Sierpiński Triangle Quilt

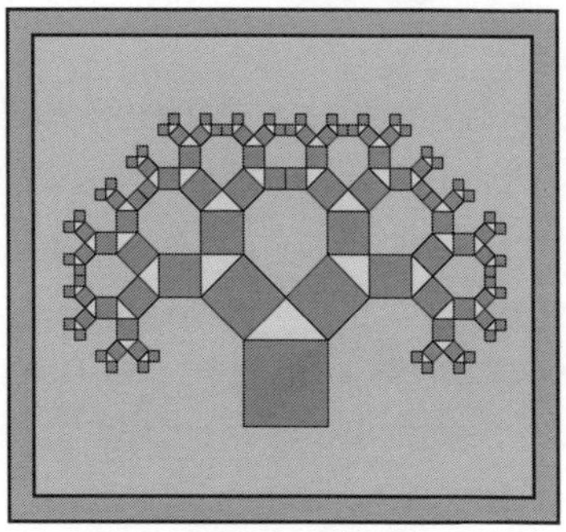

The Pythagorean Tree Quilt

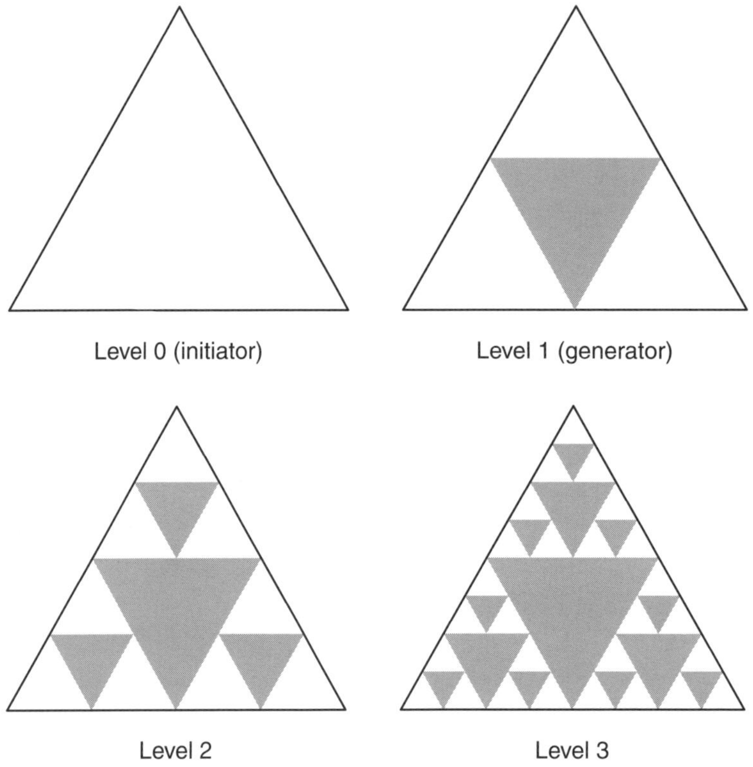

fractal. In the next iteration, the same transformation is applied to the three filled-in triangles in Level 1, creating Level 2. To make a true fractal, this process is repeated indefinitely. For the fractal design in the quilts, the generating rule is repeated five or six times.

Notice that the Sierpiński triangle is self-similar—a triangle one-fourth the size of the whole fractal is the same as the whole triangle. Parts of the fractal look like the entire fractal; each has infinitely many levels.

Mathematicians have been interested in the shapes we now call fractals for over 100 years, but until recently these shapes were regarded as mathematical curiosities, with few applications to the real world. Benoit Mandelbrot (b. 1924) introduced the term *fractal* and was the first person to recognize that many objects in nature—including trees, ferns, coastlines, and mountains—have fractal-like shapes with the whole made up of smaller parts similar to the whole.

Level 5

His work helped spark interest in fractal geometry and its applications to patterns, such as those found in weather and telecommunications.

The increasing availability of computers helped bring attention to the study of fractals. Computers allow mathematicians to create complicated fractals that they could previously only imagine. They are also able to zoom in on a computer-generated fractal to observe the details in the pattern at higher levels. Many interesting examples of fractals continue to be discovered.

The quilts studied in this chapter are based on *polygon fractals,* that is, fractals with initiators that are polygons. The Sierpiński carpet quilt begins as a square, the Sierpiński triangle as an equilateral triangle, and the Pythagorean tree as a square and an isosceles right triangle. Many polygon fractals, including the Sierpiński carpet and the Sierpiński triangle, are known as *sieves* because part of the initiator polygon is removed or "sifted out" with each iteration.

It is not possible to represent a complete fractal in a drawing or quilt because the pattern for a complete fractal repeats infinitely many times.

The Sierpiński Carpet Quilt

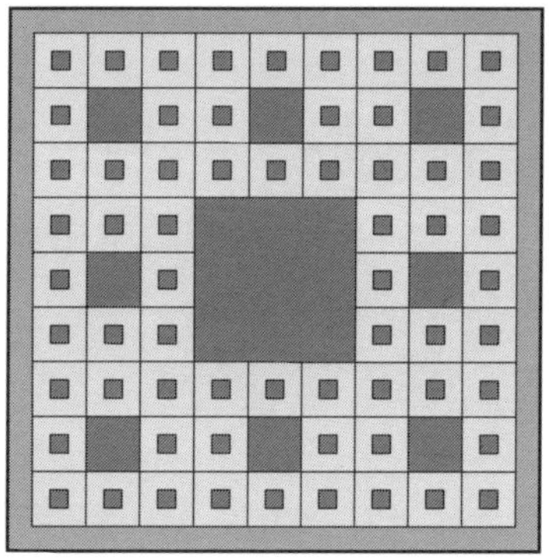

The Sierpiński carpet, also called the Sierpiński gasket, is named for Waclaw Sierpiński (1882–1969), a Polish mathematician known for his work with fractals and other self-similar shapes.

The initiator, or Level 0, for the Sierpiński carpet is a square. The generator, Level 1, can be thought of as a square divided into nine congruent squares, with the center one filled in to indicate that the square is removed. At each iteration, each remaining square is divided into nine smaller squares, and then the middle square is filled in.

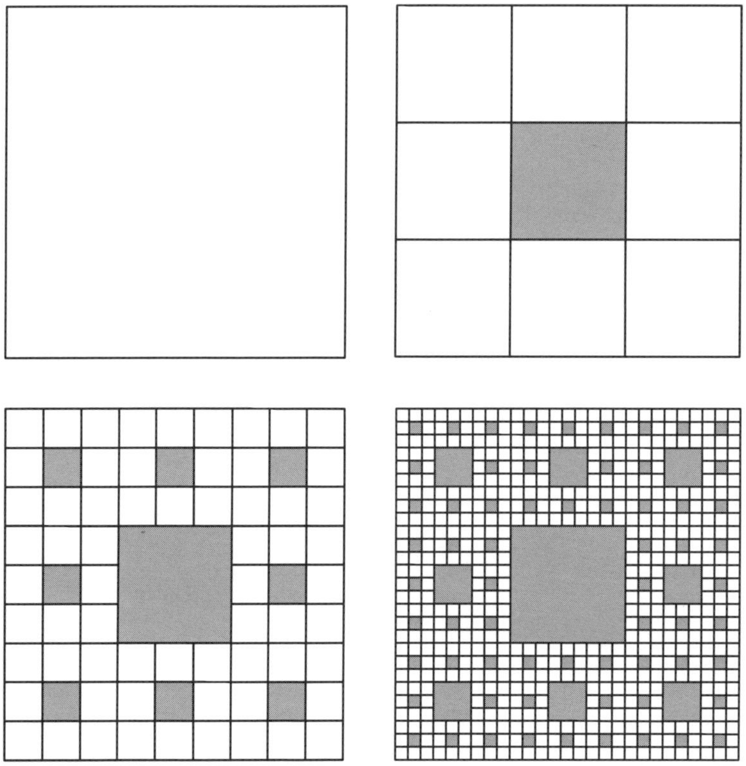

Notice that each iteration is performed within the initiator square, which becomes the outer boundary of each level.

Because the Sierpiński carpet is a simple horizontal and vertical design, the traditional colors of Pennsylvania Amish quilts were chosen for this quilt. The Amish are known for their simple lifestyle—they live off the land without using modern technology such as electricity. Their clothing is simple and is made from solid-colored fabric. Amish quilts are thought to be among the earliest art forms created by European settlers in Pennsylvania. This Sierpiński carpet quilt shows Level 3 of the fractal.

Designing Your Quilt

1. Follow these steps to create a Sierpiński carpet design.

Step 1 Draw a square on rectangular dot paper or graph paper. Because the iterations require *trisecting* the sides of each square (that is, dividing them into three equal segments), start with sides of length 3^n units so you can draw n levels of the fractal without having to use fractional side lengths. If you want to draw the figure to Level 3, start with sides of length $3^3 = 27$ units.

Step 2 Mark the points that divide each side of the square into three equal segments.

Step 3 Draw horizontal and vertical segments connecting the trisection points.

Step 4 Remove (shade) the center square.

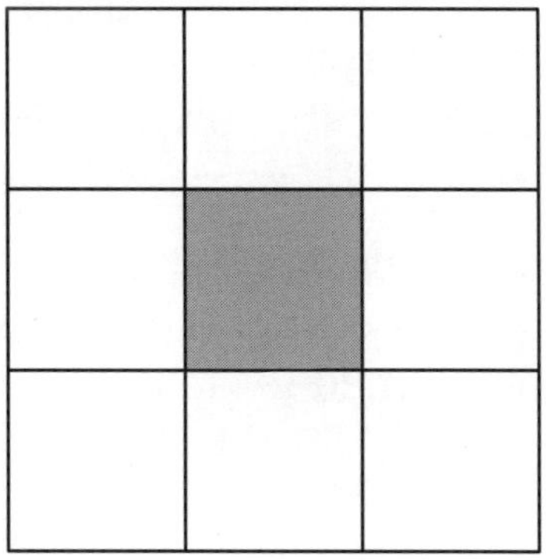

Level 1

Step 5 You have just created Level 1 of the Sierpiński carpet. This shape is known as the *generator* of the fractal because it is repeated to create, or *generate,* each level of the design. To create Level 2, follow Steps 2–4 for each of the eight remaining (unshaded) squares in Level 1.

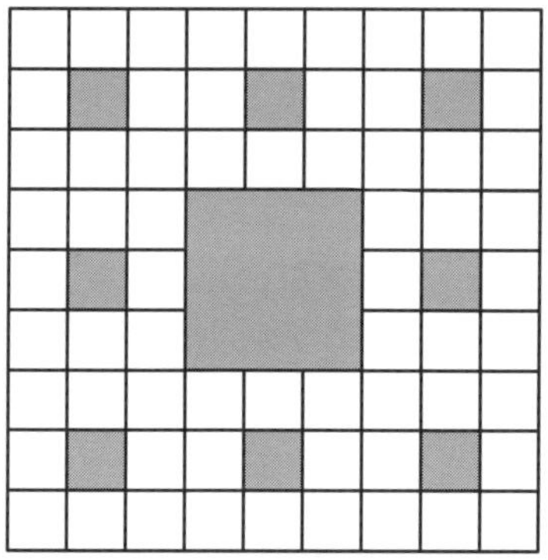

Level 2

Step 6 To draw each succeeding level, follow Steps 2–4 for each of the unshaded squares from the previous level, noting that the number of unshaded squares increases in the sequence, 8, 64, 512, 4096, ... They are the powers of 8: $8^0, 8^1, 8^2, 8^3, 8^4, \ldots, 8^n$.

Step 7 Color your design.

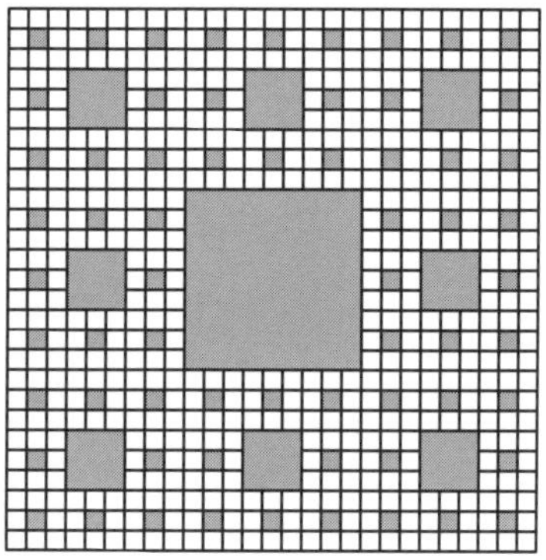

Level 3

Construct different Sierpiński carpet designs in the following activities. Try each one on a separate sheet of paper.

2. Follow the steps above to create a Sierpiński carpet design, but start by drawing a rectangle with length 27 units and width 9 units. What does this fractal look like?
3. Follow the steps above to create a Sierpiński carpet design, but instead of shading the center square in Step 3, shade a different square. Shade the corresponding square at each level. Describe the resulting design.
4. Instead of dividing the sides into three equal segments, divide them into *four* equal segments. (Start with a side length that is a power of 4.) Shade in one or more of the 16 squares formed to create the generator for a new fractal design.

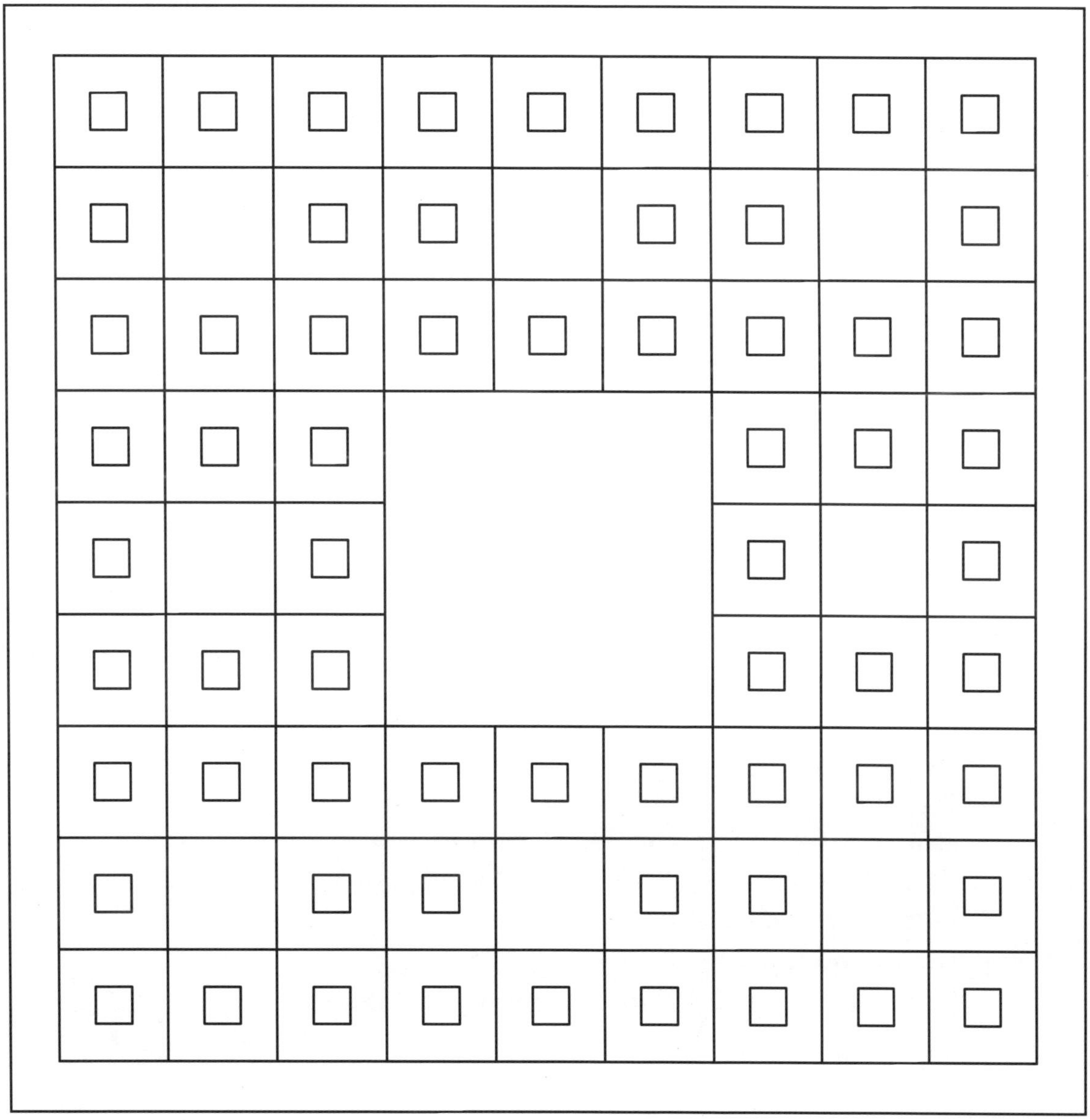

Making the Sierpiński Carpet Quilt

Materials

$\frac{1}{3}$ yard each of 8 solid-colored fabrics

1 yard of black fabric

2 yards of fabric for backing and binding

$1\frac{1}{2}$ yards of batting

Instructions

The finished size of this wall hanging is approximately 44 inches by 44 inches. Using a rotary cutter, ruler, and mat to cut fabric is recommended. A $\frac{1}{4}$-inch seam allowance is used throughout the construction of this quilt.

From each of the eight colored fabrics, cut two 2-by-22-inch strips and sixteen 2-by-5-inch strips.

From the black fabric, cut eight 2-by-22-inch strips, eight 5-inch squares, one 14-inch square, and four 4-by-44-inch strips (for the border).

Repeat the following steps for each of the eight colors:

1. Using a $\frac{1}{4}$-inch seam allowance, stitch together lengthwise three 2-by-22-inch strips: a colored strip, a black strip in the middle, and another strip of the same color. This new strip of three fabrics should be 5 by 22 inches. Press the seams to the black fabric. From this new strip of three fabrics, cut eight 2-by-5-inch strips. Each of these strips will be the center strip for one of the smaller squares in the design.

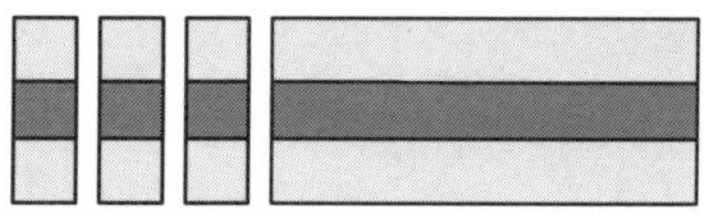

2. Use the sixteen 2-by-5-inch strips of colored fabric as the top and bottom strips for one of the smaller squares in the design.

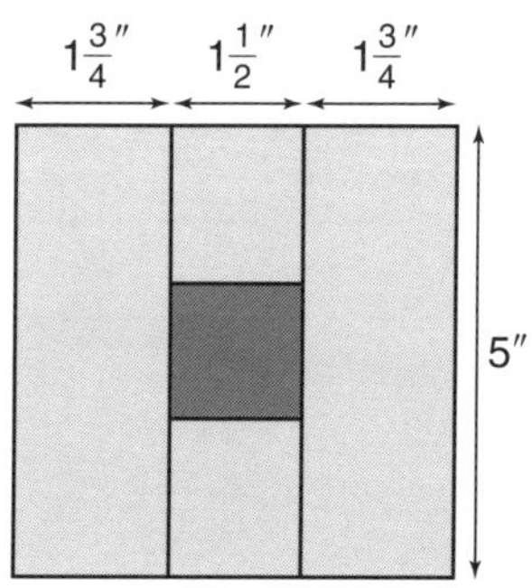

3. Create eight of the smaller squares by stitching three of the strips from the previous steps. The new square should measure 5 inches on each side. Each of these squares is Level 1 of Sierpiński's carpet.

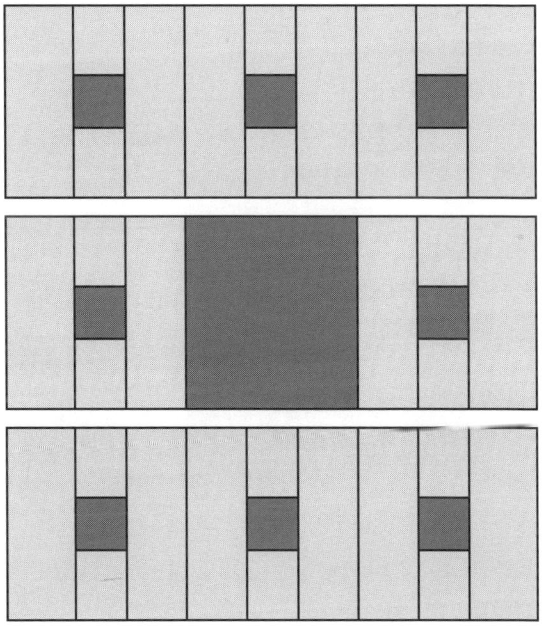

Repeat the preceding steps for the remaining seven colored fabrics.

For each of the eight colors, arrange its eight Level 1 squares around a 5-inch center black square. Look at the arrangements and decide if rotating the squares to alternate the grain of the fabric would be more interesting. Stitch these nine squares together by first stitching the three squares in each row together, and then stitching the three rows together. Each of these squares is Level 2 of Sierpiński's carpet.

Arrange the eight Level 2 squares around the 14-inch square of black fabric. Sew together in rows as in the previous step. The design of the Sierpiński's carpet quilt is complete.

To complete the wall hanging, add a border of black (1–2 inches wide) around all four sides of the design. Layer the quilt top, batting, and backing, and baste the layers together. Quilt to outline the smaller squares. Add a binding to the outer edge. To hang the quilt, stitch a tubular cloth sleeve to the top back of the quilt and insert a wooden rod into the sleeve.

The Sierpiński Triangle Quilt

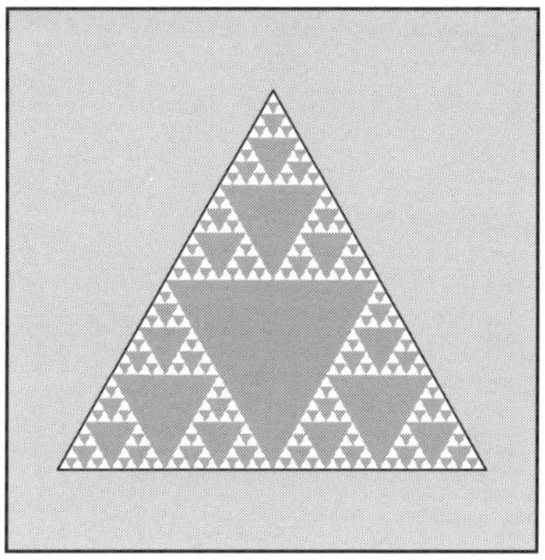

The Sierpiński triangle is a popular fractal design found in many old and new forms of art. The initiator for the fractal is a triangle. In both quilts shown, an equilateral triangle was used, but other triangle shapes may be used as well.

The transformations used to create the Sierpiński triangle removed a triangular piece; similarly, the Sierpiński carpet removed a square. At each iteration, the midpoints of the sides of each remaining triangle are connected, forming four identical triangles, and then the center triangle is filled in to indicate that it has been removed.

Notice that each iteration is performed within the initiator triangle, so this triangle is the outer boundary of each level.

The red-bordered quilt (top right) shows Level 4 of the Sierpiński triangle. In this quilt, yellow (shown in light gray) represents the removed

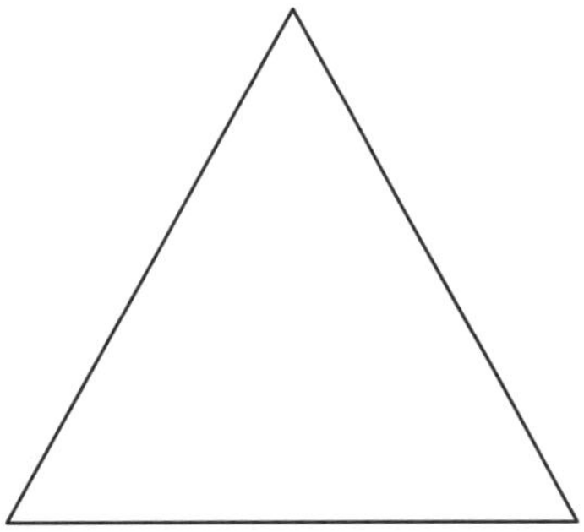

Level 0 (initiator)

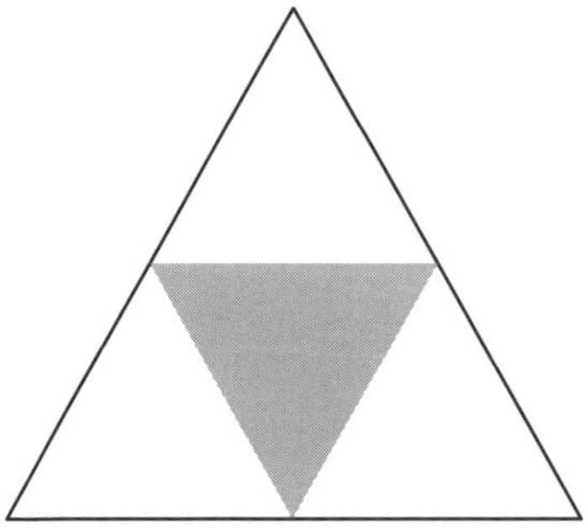

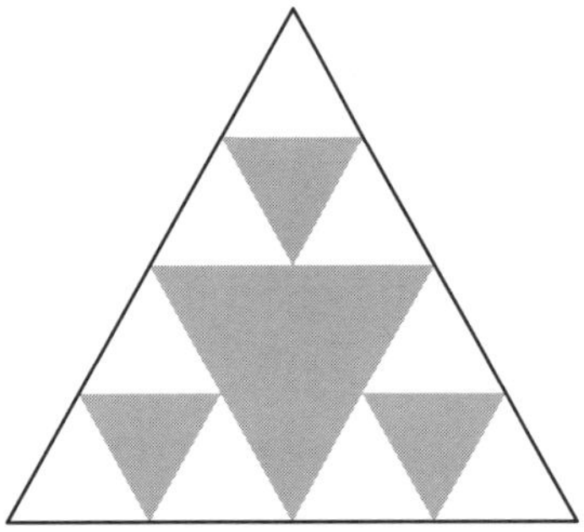

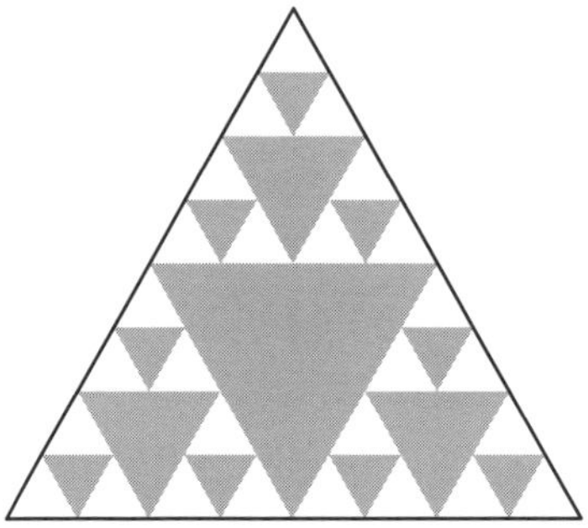

portions of the design, and 81 different fabrics represent the remaining triangles. The other quilt represents Level 5 of the fractal design, with dark blue (shown in gray on previous page) representing the removed areas and other solid colors representing the remaining triangles. Like Amish quilts, this quilt uses solid colors dominated by dark colors such as navy blue or black.

Designing Your Quilt

1. Follow these steps to create a Sierpiński triangle design.

Step 1 On a sheet of isometric dot paper, draw an equilateral triangle. Because each iteration requires bisecting the sides, choose a side length that is a power of 2. To draw the fractal to Level 4, use side lengths of 2^4, or 16, units.

Step 2 Mark the midpoint of each side of the triangle.

Step 3 Using a straightedge, connect the three midpoints. The figure should now contain four identical triangles.

Step 4 Remove (shade) the center triangle.

Step 5 You have now created Level 1 of the Sierpiński triangle. This shape is the *generator*, which is repeated to create each level of the design. To create Level 2, follow Steps 2–4 for each remaining (unshaded) triangle in Level 1.

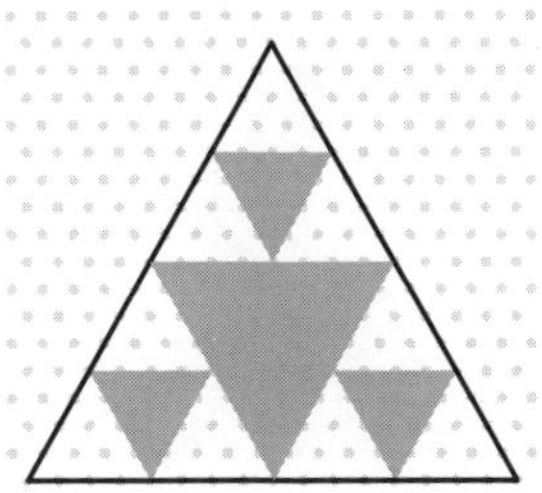

Level 2

Step 6 To draw each succeeding level, follow Steps 2–4 for each unshaded triangle from the previous level.

Level 3

Level 4

Step 7 Color your design.

2. Follow the steps in part 1, but start with an isosceles right triangle or a scalene triangle instead of an equilateral triangle. (Use rectangular dot paper or graph paper.)

WORKSHEET 2

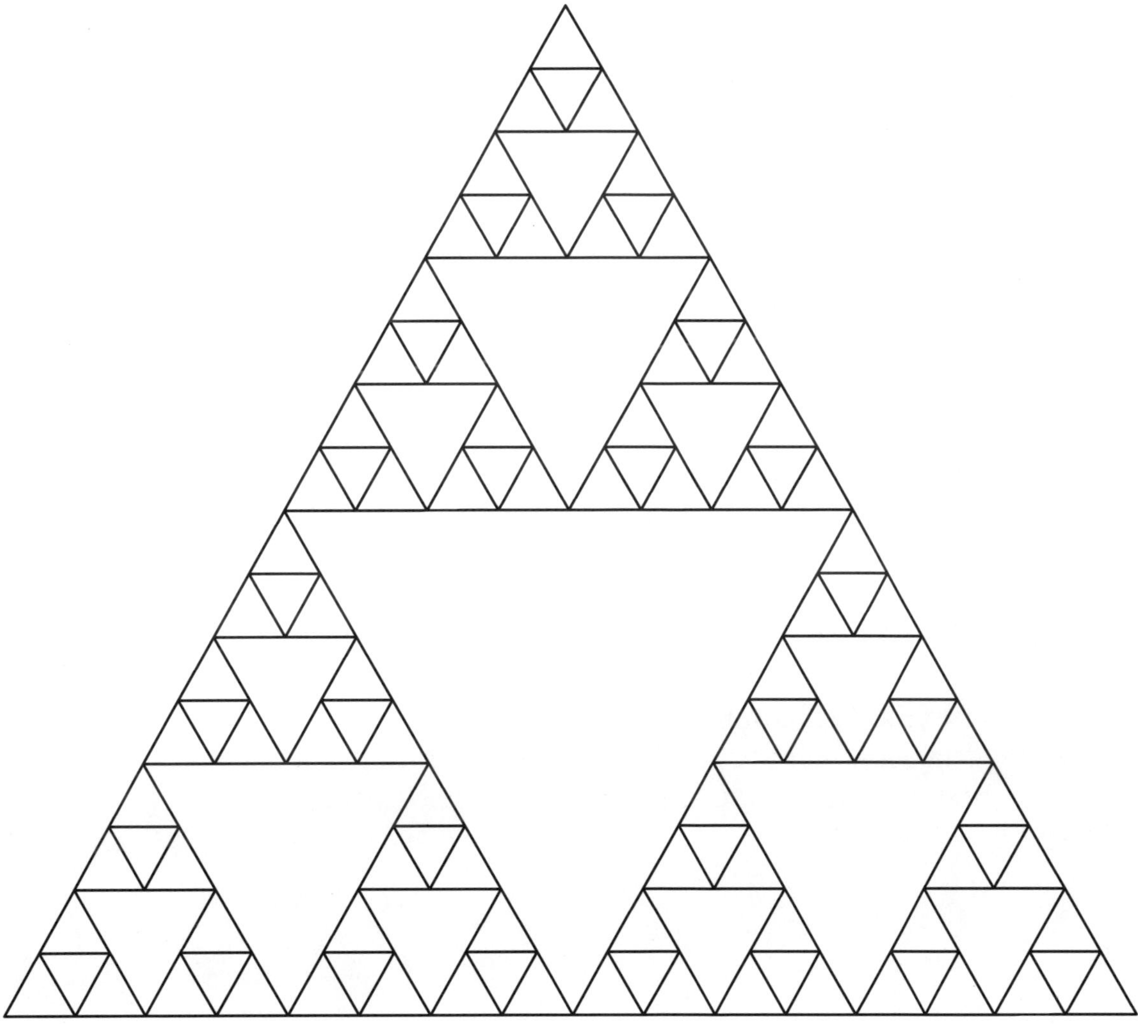

Making the Sierpiński Triangle Quilt

These instructions are for Level 4 of the Sierpiński triangle. The finished quilt should measure 34 inches in height and width.

Materials

$1\frac{1}{2}$ yards of fabric for background and backing; add another $\frac{1}{2}$ yard for binding

One 3-inch square of fabric is needed for each of the 81 colored triangles in Level 4 of this design; use 81 different fabrics or a variety of your own design

$1\frac{1}{4}$ yards of batting

If a rectangular quilt is desired, add 1 yard of background fabric for additional triangles needed to square the corners of the finished quilt

Instructions

Cut a 34-inch square from the background fabric and set aside. If you are using this fabric for binding, cut three 24-by-44-inch strips and set aside. Using triangles *A* and *B*, make a cardboard or plastic template of each for the smaller triangles in the quilt. Cut out the templates and label them *A* and *B*. Use triangle *A* to trace 81 triangles on the back of colored fabrics and 27 triangles on the back of the background fabric. These will be the cutting lines for the triangles. Before cutting the triangles, trace triangle *B* in the center of each triangle *A*. These will be the stitching lines for the triangles.

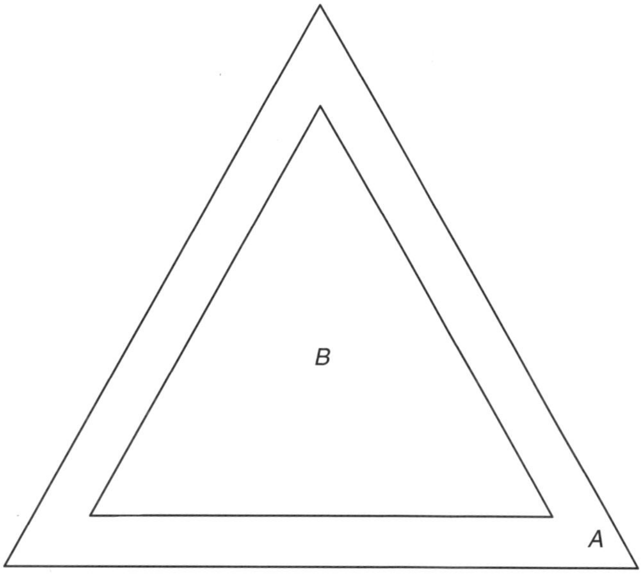

At this point, there should be 81 colored triangles and 27 background triangles. Arrange these triangles in 27 groups of four, with each group having three colored triangles and one background fabric. Sew three of the colored triangles to the background triangle, one to each side, using the pencil line as a guide. Press the seams toward the colored triangles. Each of these triangles is Level 1 of the Sierpiński triangle.

Using one of these triangles as a pattern for triangle *C*, make a template and trace nine triangle *C*'s on the back of the background fabric, add a seam allowance, and cut them out. Arrange three of the Level 1 triangles from the previous steps around one triangle *C* and stitch them together, using the pencil lines from triangle *B* as a guide. Press the seam allowance away from the background triangle. Each of these nine triangles is Level 2 of Sierpiński's triangle.

Using one of these Level 2 triangles as a guide, make a template of triangle *D* and use it to trace three triangle *D*'s on the back of the background fabric. Add a seam allowance and cut these out and arrange three of the previous Level 2 triangles around each one. Stitch them together with triangle *D* in the center and then press. Each of these three triangles is Level 3 of Sierpiński's triangle.

Using one of the Level 3 triangles, make a template for triangle *E* and trace one triangle *E* on the background fabric. Again add a seam allowance before you cut it out. Arrange the three Level 3 triangles around triangle *E* and stitch them together following the triangle *B* lines and press. Level 4 of Sierpiński's triangle is complete. (At this point, if a rectangular quilt is desired, add triangles of background fabric to two sides of the equilateral triangle quilt to form a rectangle.)

To add a border around the perimeter of the Sierpiński's triangle, cut strips of the desired width and stitch to the three sides. Pin or baste the completed triangle to the batting and backing, and quilt the layers together. Add a binding. To hang a triangular quilt, hook-and-loop fasteners can be applied to the back of the quilt and to a piece of foam board cut slightly smaller than the quilt.

The Pythagorean Tree Quilt

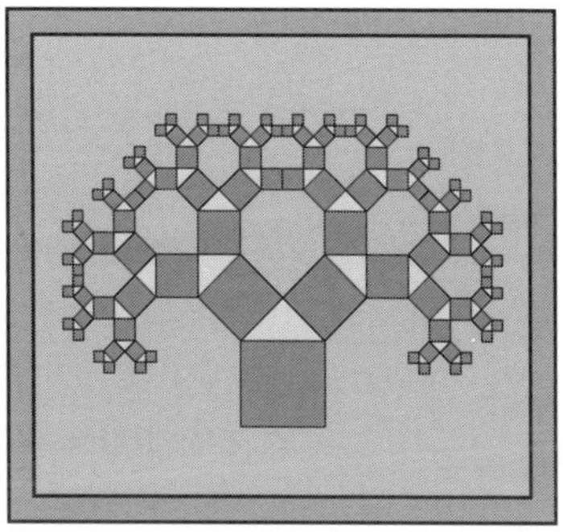

The Pythagorean tree fractal contains a visual proof of the Pythagorean theorem, and as it continues to advanced levels, a tree shape is formed. The initiator shape for the fractal consists of a rectangle and a right triangle whose hypotenuse is the length of one side of the rectangle. For the quilt, a square and an isosceles right triangle were used. The iterations of the fractal result from building a square upon the legs of the triangles from the previous level and adding a triangle to the opposite side of each square.

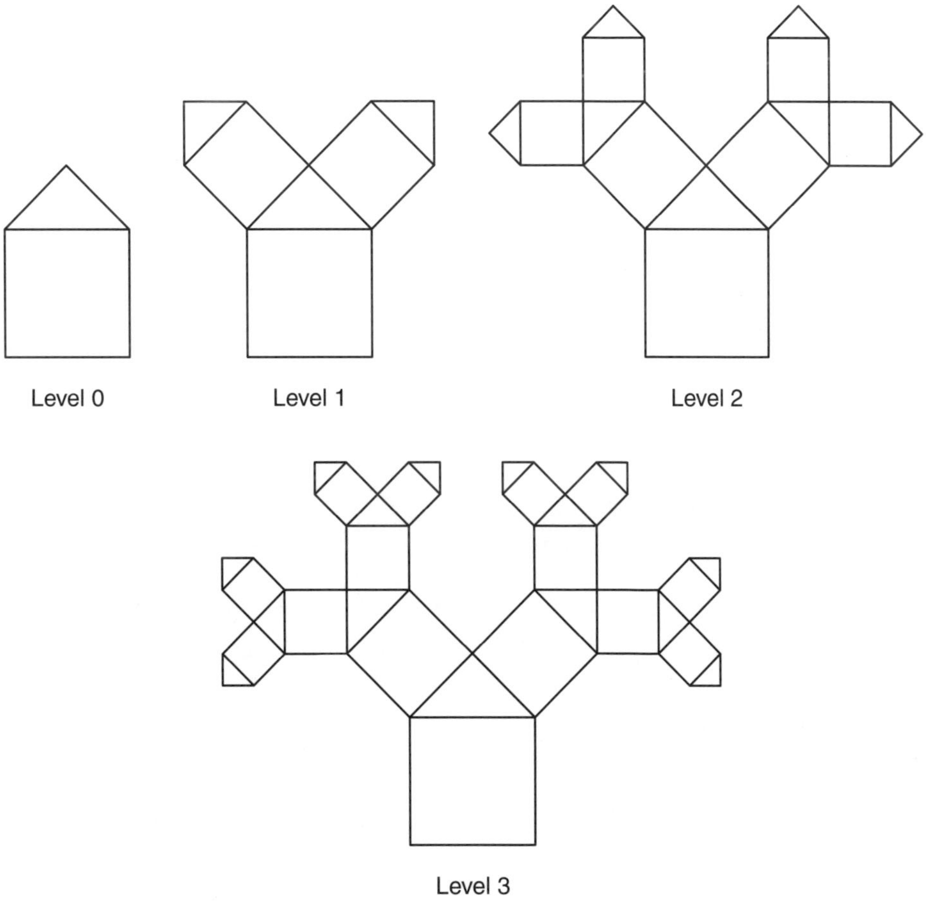

While the Sierpiński carpet and Sierpiński triangle fractals (see pages 4 and 10) are entirely contained within their initiator shape, the Pythagorean tree grows outward, resulting in a beautifully symmetric tree shape. The tree is symmetric because the square and triangle are symmetric. As this fractal tree shape suggests, many natural objects—such as broccoli stalks, fern leaves, coastlines, and mountains—have fractal-like constructions.

A. E. Bosman (1901–1961), a German mathematician and World War II submarine designer, was the first to construct the fractal tree with isosceles triangles and squares. In his published work, Bosman expressed his wonder at the miraculous geometrical shapes of nature.

The Pythagorean tree quilt shows Level 5 of the fractal. A tan fabric was used to represent the trunk and branches of the tree. A green fabric with a floral pattern was chosen to represent the leaves and colored fruit or berries on the tree.

Designing Your Quilt

You will need a large sheet of rectangular dot paper or graph paper to make the Pythagorean tree design. To create Level 5, the grid needs to be at least 80 by 56 units. You may need to tape together several sheets of paper.

1. Follow these steps to make the design.

   Step 1 Turn the paper so that the long side is at the bottom. Draw a square with side length 16 units centered near the bottom of the paper.

   Step 2 Draw an isosceles right triangle with the top of the square as its hypotenuse. To do this, first draw a vertical segment half the length of the square's side, extending up from the midpoint of the top of the square. Form the triangle by connecting the top of this segment to the top two vertices of the square.

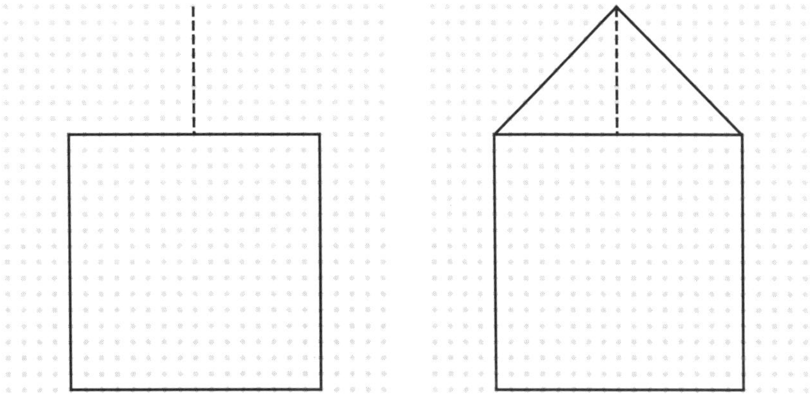

The resulting shape is the initiator of the fractal. To create each successive level of this fractal design, repeat the following steps:

Step 3 Draw a square on each leg of the triangles from the previous level. Use the horizontal, vertical, and diagonal units on the paper as a guide rather than actually measuring the sides.

Step 4 Draw an isosceles right triangle on each new square.

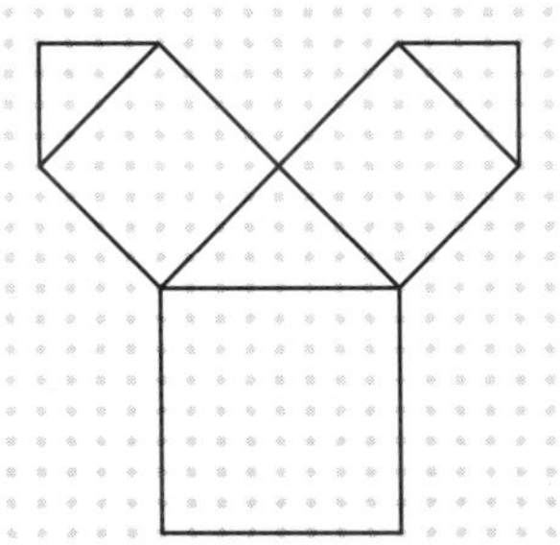

Level 1

Step 5 Create five or more levels, then color your design.

2. Begin with a rectangle that is not a square and use triangles whose legs are the lengths of the sides of the rectangles.

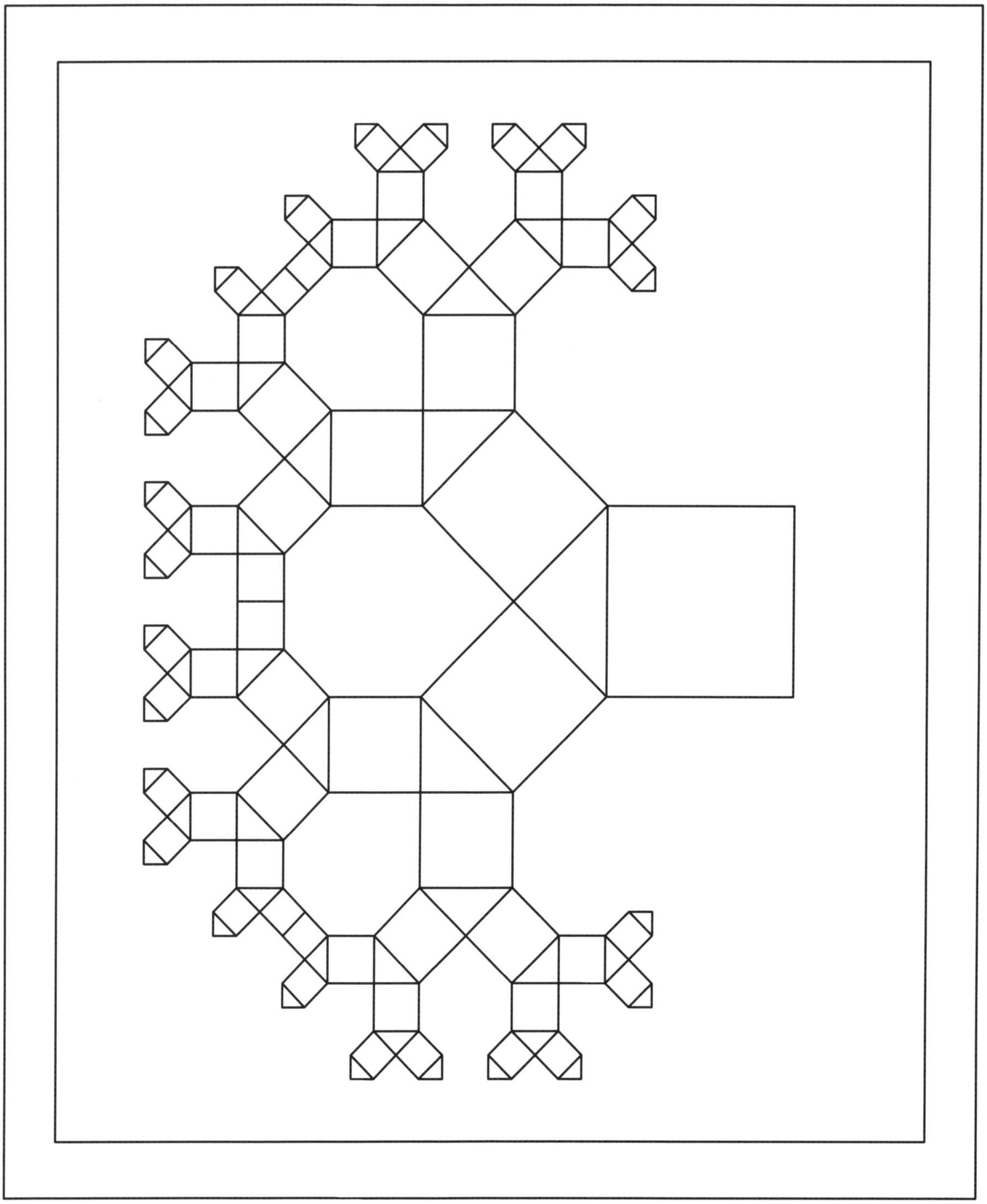

Making the Pythagorean Tree Quilt

The completed quilt measures 25 by 36 inches, including borders.

Materials

$\frac{1}{4}$ yard of fabric for squares that represent the trunk and branches

$\frac{1}{4}$ yard of fabric for triangles that represent the leaves

$\frac{3}{4}$ yard of background fabric

$\frac{3}{4}$ yard of backing fabric

$\frac{1}{8}$ yard and $\frac{1}{4}$ yard of fabric for the two borders

$\frac{1}{4}$ yard of binding fabric

$\frac{3}{4}$ yard of batting

Instructions

This quilt is created using English paper piecing, a hand-sewing method in which paper shapes are cut and basted to the fabric shapes before the quilt is assembled. This technique preserves the shapes of the pieces and eliminates fabric stretching along the bias edges of the triangles.

Draw a full-size Pythagorean tree to Level 5, following the steps in the "Designing Your Quilt" section. For the quilt shown, the tree was drafted on four sheets of 13-by-22-inch quarter-inch graph paper taped together. The first square measured 5 by 5 inches (that is, 20 by 20 units).

Label same-size pieces with the same number or letter. For example, label all the smallest squares 1, all the smallest triangles 2, and so on. Cut the pieces apart. Lay each paper piece on the fabric and cut a corresponding fabric piece, adding a $\frac{1}{4}$-inch seam allowance on all sides. Then, folding the seam allowances to the back of the paper, hand baste each fabric piece to its corresponding paper piece, with the right side of the fabric facing out.

Leaving the paper pieces intact, construct the tree by joining appropriate pieces, edge to edge, on the right side of the fabric, using small blind overcast stitches. It is easiest to make the tree in sections and then join the sections.

Once the tree is completed, lay it on the background fabric, with the paper still intact, allowing several inches around the tree. Pin or baste the tree to the background, making sure that the symmetry is preserved and that the left and right branches of the tree are aligned.

Appliqué the tree to the background. Then cut away the background fabric from behind the tree, leaving a $\frac{1}{4}$-inch seam allowance. Carefully remove the basting and paper from the tree. Press the tree and background fabric to remove any wrinkles. Add the two borders to the background fabric.

Baste the finished tree to the batting and backing with thread or small safety pins. Then quilt the layers together by outlining the tree, adding any embellishments desired. Bind the quilt and add a rod pocket for hanging.

Exploring Area and Perimeter of the Sierpiński Carpet

This Level 3 Sierpiński carpet has an initiator square with side length 27 units.

1. Find the area and perimeter of each level from Level 0 (the initiator square) through Level 3.

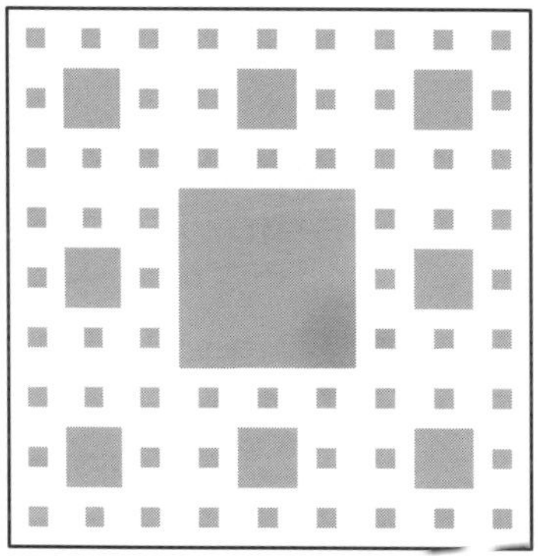

Level 3

	Area Removed (square units)	Area Remaining (square units)	Perimeter (including perimeters of removed squares, or units)
Level 0			
Level 1			
Level 2			
Level 3			

2. Describe the patterns you see in the areas and perimeters as the design changes from Level 0 to Level 3.
3. Calculate the following ratios:

$\frac{\text{area of Level 1}}{\text{area of Level 0}}$

$\frac{\text{area of Level 2}}{\text{area of Level 1}}$

$\frac{\text{area of Level 3}}{\text{area of Level 2}}$

Explain the results.

Exploring Sierpiński Triangle Areas

Here is Level 4 of the Sierpiński triangle.

1. When you draw this fractal design, it is a good idea to begin with a triangle whose side lengths are a power of 2. Explain why. To draw Level 4, how many units on a side should the initial triangle be so that the smallest Level 4 triangle is one on a side?
2. Suppose the area of Level 0 (the initiator triangle) is 1 square unit.
   a. Express the area of each level in both fraction and decimal form, and indicate the area that has been removed.

	Area in Fraction Form (square units)	Area in Decimal Form (square units)	Total Area Removed (square units)
Level 0			
Level 1			
Level 2			
Level 3			
Level 4			

   b. As you move through the levels, do the areas get larger or smaller?
   c. Describe the patterns you see in the fractions and decimals, and try to explain why these patterns occur.
   d. Express the areas in question 2a as powers of the same fraction.
   e. Predict the areas for Levels 5 and 6 using the patterns from the table. Do these decimals seem to be approaching a number? Will they ever reach that number? Explain.

Exploring Areas of a Pythagorean Tree

This Pythagorean tree design has an initiator square with side length 16 units and an isosceles right triangle with hypotenuse 16 units.

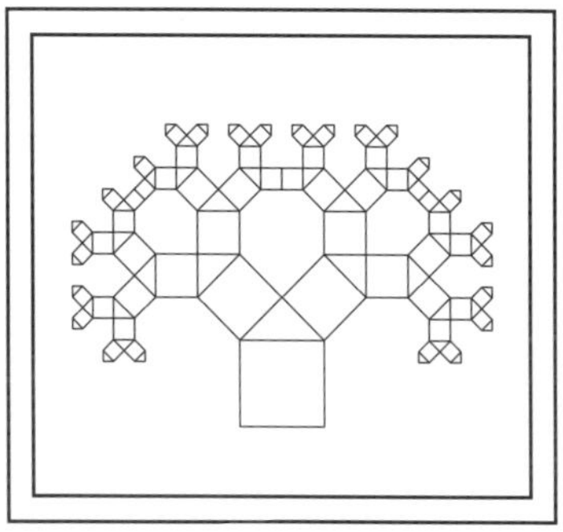

Complete the table.

	Level 0 (initiator)	Level 1	Level 2	Level 3	Level 4	Level 5
Total number of squares	1					
Area of each new square	256					
Total area of all squares	256					
Total number of triangles	1					
Area of each new triangle	64					
Total area of all triangles	64					
Total area of design	320					

1. For which squares and triangles was it easiest to find the area?
2. What is the relationship between the area of each triangle and the area of the square on which it is constructed?
3. Describe the patterns you observe in the areas of the squares.
4. Describe the patterns you observe in the areas of the triangles.
5. As the areas of the new triangles and squares get smaller, what happens to the total area? Why?

The Sieve of Eratosthenes

The *sieve of Eratosthenes* is a number sieve that starts with the set of natural numbers (1, 2, 3, 4, . . .) and systematically removes all the composite numbers, leaving only prime numbers.

1. Create a sieve of Eratosthenes using the grid on page 27. Which of the numbers from 1 through 60 are prime?
2. Describe the pattern in the multiples of 2. In which columns do they appear? What fraction of the numbers are removed when you sift out the multiples of 2?
3. Describe the patterns in the multiples of 3, of 5, of 7, and of 11.
4. Which of the patterns you describe in questions 2 and 3 occur because the sieve is arranged in six columns?
5. If the sieve were arranged in ten columns instead of six, different patterns would occur. How many columns would contain the multiples of 2? How would the patterns for the multiples of 3 and multiples of 5 change?
6. Use the sieve to list the prime factors of each number in this table. (The prime factors are circled numbers with a marking also on the composite.) Then write the number as the product of primes.

Number	Prime Factors	Product of Primes
48		
64		
51		
60		
57		

7. Twin primes are primes that are consecutive odd numbers, such as 3 and 5. Identify the twin primes in your sieve.

The Sieve of Eratosthenes (continued)

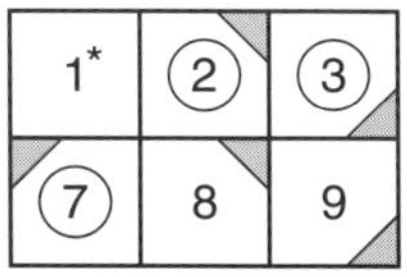

Put an asterisk (*) next to the 1 to indicate that it is neither prime nor composite.

Circle the 2 to indicate that it is prime. Then mark the square in some way and make the same mark in the square for each multiple of 2 (that is, 4, 6, 8, 10, and so on).

Move to the next unmarked square. It is prime. Circle it and mark the grid square in a unique way. Make the same mark in the square for each multiple of that number and so on.

1	2	3	4	5	6
7	8	9	10	11	12
13	14	15	16	17	18
19	20	21	22	23	24
25	26	27	28	29	30
31	32	33	34	35	36
37	38	39	40	41	42
43	44	45	46	47	48
49	50	51	52	53	54
55	56	57	58	59	60

ACTIVITY 5

Squares on the Sides of a Right Triangle

1. On grid paper, draw the initiator for the Pythagorean tree—a square and an isosceles right triangle. Then construct squares on the two legs of the right triangle.
   a. Find the area of the large square and the area of each of the smaller squares.
   b. Describe the relationship among the areas you found in 1a.
2. Repeat part 1, but start with a square of a different size. Compare your findings with your findings from the first drawing and with the findings of your classmates.
3. Make a conjecture about the areas of squares drawn on the sides of an isosceles right triangle.
4. Repeat the investigation of areas of squares on the sides using a right triangle that is not isosceles. To find a right angle for a nonisosceles triangle, use a compass to draw a semicircle with the side of the square as a diameter. Any point on that semicircle is the vertex of a right triangle with hypotenuse the side of the square.
5. Make a conjecture about the areas of squares drawn on the sides of a right triangle.
6. Compare your conjecture to the statement of the Pythagorean theorem: The sum of the squares of the lengths of the legs of a right triangle equals the square of the length of the hypotenuse.

ACTIVITY 6

The Chaos Game

To play the Chaos Game, you need Worksheet 4, "The Chaos Gameboard." The worksheet shows a large equilateral triangle with vertices labeled *T* (top), *L* (left), and *R* (right). You also need a ruler and a device for randomly selecting a vertex of the triangle (for example, a spinner with three equal sections, or a die for which two numbers correspond to each vertex).

1. Mark a point anywhere inside or outside the triangle. Then follow these steps:

   Step 1 Randomly select vertex *T*, *L*, or *R* using a spinner, die, or some other device.

   Step 2 Plot a point halfway between the last point you marked and the vertex you selected in Step 1.

   Repeat these steps, always measuring from the last point you marked, until you see a pattern emerge. (You may need to plot fifty or more points.) What do you think the results would look like if you plotted thousands of points?

2. If you and your classmates played the game on transparencies of the worksheet, lay the transparencies on top of each other, lining up the original triangles, and display the results on an overhead projector.

3. Suppose you had started the Chaos Game with an isosceles right triangle. Sketch what you think the results would look like.

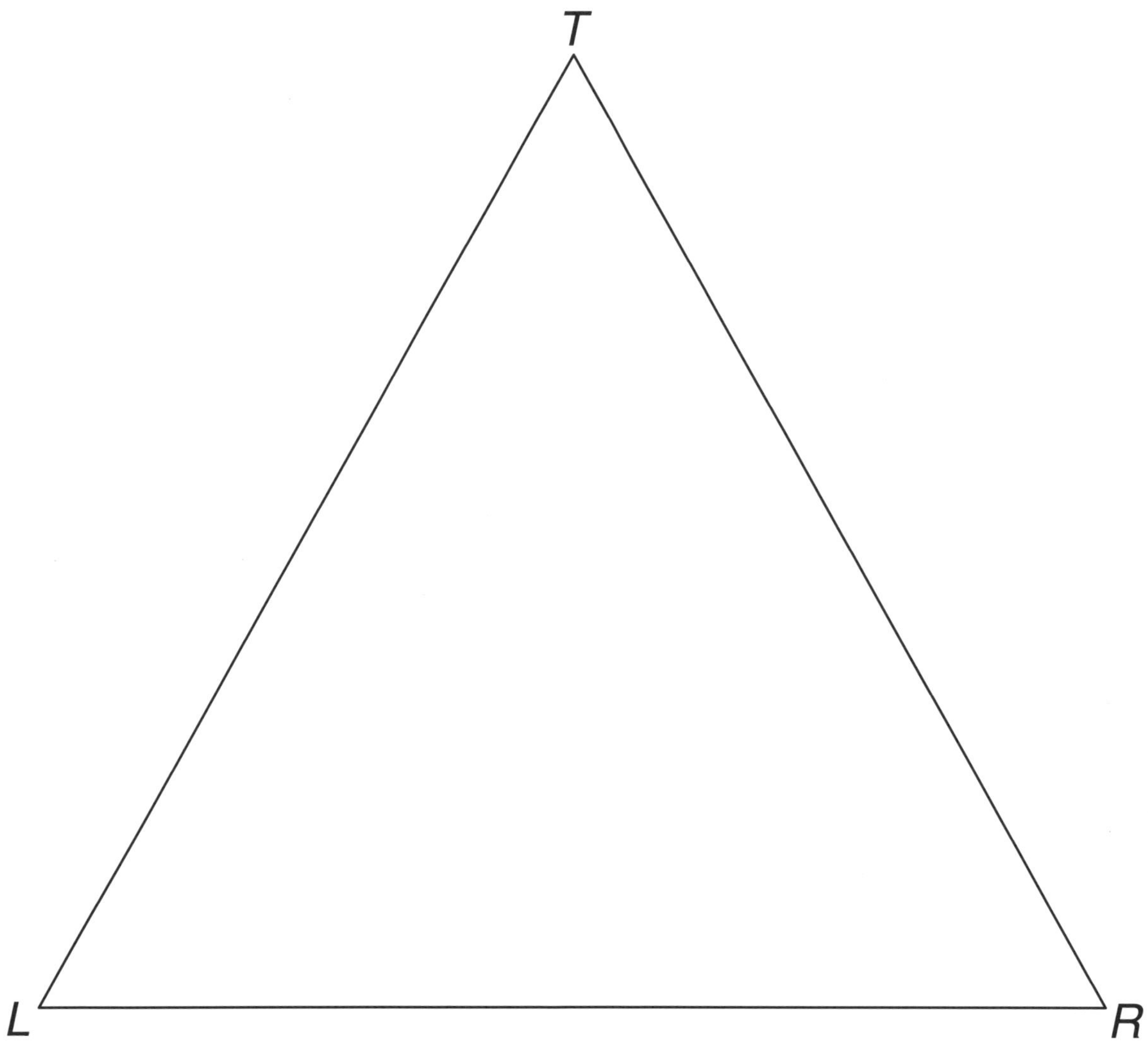
T
L
R

Graphing Calculator Exploration

Enter the programs on a TI-73, TI-83, or TI-83 Plus calculator to play the Chaos Game. To generate the fractal from the quilt design, choose the equilateral triangle with the fraction $\frac{1}{2}$.

TI-73

```
Program:CHAOS
ClrScreen
Menu("CHAOS SHAPES","RIGHT
  TRIANGLE",1,"EQUILATERAL TRI",2,
  "RANDOM TRIANGL",3,"SQUARE",4,
  "PENTAGON",5,"HEXAGON",6,"QUIT",8)
Lbl 1
Disp "RIGHT TRIANGLE"
{.05,.95,.05}→L1
{.05,.05,.65}→L2
Goto 7
Lbl 2
Disp "EQUILATERAL TRIA"
{.05,.95,.5}→L1
{.05,.05,.65}→L2
Goto 7
Lbl 3
Disp "RANDOM TRIANGLE"
{rand ,rand ,rand }→L1
.7{rand ,rand ,rand }→L2
Goto 7
Lbl 4
Disp "   SQUARE"
{.2,.8,.8,.2}→L1
{.05,.05,.65,.65}→L2
Goto 7
Lbl 5
Disp "   PENTAGON"
.5+.35sin({0,1,2,3,4}72°)→L1
.35+.35cos({0,1,2,3,4}72°)→L2
Goto 7
Lbl 6
Disp "   HEXAGON"
.5+.35sin({0,1,2,3,4,5}60°)→L1
.35+.35cos({0,1,2,3,4,5}60°)→L2
Lbl 7
ClrDraw:FnOff
PlotsOff
0→Xmin:1→Xmax
0→Ymin:.7→Ymax
Plot1(xyLine,L1,L2,▫)
L1(1)→X:L2(1)→Y
dim(L1)→N
Disp "","   ENTER A"
Disp "  FRACTION"
Disp "BETWEEN 0 AND 1"
Input F
For(J,1,1000)
int(Nrand +1)→P
X+(L1(P)-X)F→X
Y+(L2(P)-Y)F→Y
Pt-On(X,Y)
End
Text(0,84,"P E"
Text(6,84,"R N"
Text(12,84,"E T"
Text(18,84,"S E"
Text(24,84,"S R"
Pause
Lbl 8
PlotsOff
Disp "  PRESS", "  ENTER TWICE","  TO
  REPLAY","","  PRESS","ENTER, 1,
  ENTER","  TO QUIT"
Pause
```

TI-83/TI-83 Plus

```
Program:CHAOS
ClrHome
Menu("CHAOS SHAPES","RIGHT
  TRIANGLE",1,"EQUILATERAL TR",2,
  "RANDOM TRI",3,"SQUARE",4,
  "PENTAGON", 5, "HEXAGON",6,
  "QUIT",8)
Lbl 1
Disp " RIGHT TRIANGLE"
{.05,.95,.05}→L1
{.05,.05,.65}→L2
Goto 7
Lbl 2
Disp "EQUILATERAL TRIANGLE"
{.05,.95,.5}→L1
{.05,.05,.65}→L2
Goto 7
Lbl 3
Disp "RANDOM TRIANGLE"
{rand ,rand ,rand }→L1
.7{rand ,rand ,rand }→L2
Goto 7
Lbl 4
Disp "    SQUARE"
{.2,.8,.8,.2}→L1
{.05,.05,.65,.65}→L2
Goto 7
Lbl 5
Disp "    PENTAGON"
.5+.35sin(({0,1,2,3,4}72°)→L1
.35+.35cos(({0,1,2,3,4}72°)→L2
Goto 7
Lbl 6
Disp "    HEXAGON"
.5+.35sin(({0,1,2,3,4,5}60°)→L1
.35+.35cos(({0,1,2,3,4,5}60°)→L2
Lbl 7
ClrDraw:FnOff
PlotsOff
0→Xmin:1→Xmax
0→Ymin:.7→Ymax
Plot1(Scatter,L1,L2,□)
L1(1)→X:L2(1)→Y
dim((L1)→N
Disp "","    ENTER A"
Disp "    FRACTION"
Disp "BETWEEN 0 AND 1."
Input "F=",F
For(J,1,1000)
(int(Nrand +1)→P
X+(L1(P)-X)F→X
Y+(L2(P)-Y)F→Y
Pt-On(X,Y)
End
Text(0,84,"P E")
Text(6,84,"A N")
Text(12,84,"E T")
Text(18,84,"S E")
Text(24,84,"S R")
Pause
Lbl 8
PlotsOff
Disp "  PRESS ENTER",
  "TO REPLAY","","    PRESS",
  "  1 AND ENTER","TO QUIT"
```

Activity 8

Research Topics

Research any topic listed below. You might search the Internet to find information about fractals and other topics in this chapter. (Note: Using "AND" between two terms finds only those pages that mention both terms.)

Bosman, A. E.
chaos
Eratosthenes
fractal
fractal dimension
Minkowski, Hermann
Pascal, Blaise
Pythagoras
Pythagorean Society
Pythagorean tree
Sierpiński carpet
Sierpiński gasket
Sierpiński, Waclaw
sieves in mathematics
von Koch, Helge

TEACHER NOTES

Most students find studying fractals to be interesting and exciting. As they draw the designs on dot paper or graph paper, students can observe how the fractal becomes more complex.

Most of these fractal designs can be modified—the original shape or transformation can be changed—resulting in new fractal shapes. As students create variations, encourage them to keep the changes simple, thinking ahead to higher levels where the transformations must be done many times. When students have completed their designs, they can add color to their creations. Encourage them to be creative.

The fractals in this chapter are called polygon fractals because their initial shape is formed by one or more polygons. Sierpiński's carpet begins with a square, and its transformations are easy to recreate. Sierpiński's triangle is slightly more difficult; using isometric dot paper makes it easier to draw. These two fractal designs grow inward. Thus, the first square or triangle must be large enough to allow for repeated subdivisions.

For Sierpiński's carpet, trisections are used. Talk with students about powers of 3, and help them use powers of 3 for the number of units on a side. Multiples of 3 are not as helpful. For example, 12 would work initially but would then require fractions in the units resulting from trisecting a square with 4 units to a side. Powers of 3 will allow for repeated trisections. Sierpiński's triangle is created with bisections. Talk with students about powers of 2. For a Level 4 design, 32 or 16 is a good number of units on a side of the initiator triangle.

Sierpiński's carpet or triangle can also be generated with the Chaos Game. This game can be played manually (Activity 6) or on graphing calculators (Activity 7).

The third polygon fractal, the Pythagorean tree, is a good fractal for introducing the idea of fractals in nature. Making the tree with right triangles and squares provides an opportunity to make a connection with the Pythagorean theorem, because a visual representation of the theorem is embedded in the tree. Students can use graph paper to count squares and demonstrate the theorem for themselves. In nature, trees are not symmetric because of the variance in natural events—rainfall, sunlight, wind, soil conditions, and changes in the genetic structure of seeds. Students can create a nonsymmetrical Pythagorean fractal tree with a rectangle and a right triangle whose legs are the lengths of the sides of the rectangle.

Before students work on the quilt designs or the activities, you might discuss with them the characteristics of fractals and the history of fractal geometry (see pages 1–3). Introduce the terms *initiator*, *iteration*, and *generator*, and illustrate these ideas with an example.

The activities in this chapter encourage the exploration of patterns within the fractal designs and extend mathematical ideas related to the fractals. Dot paper or graph paper is useful for many of the activities. In addition to helping students draw neat, regular patterns, a grid allows them to determine areas of shapes by counting square units. This is an important step in developing the concept of area, and it aids in understanding the areas found by using formulas.

In Activities 1, 2, and 3, students explore patterns in the three fractal designs used in the quilts: the Sierpiński carpet, the Sierpiński triangle, and the Pythagorean tree. Pattern exploration is an important part of mathematics. As students investigate successive levels of the fractals, they are asked to observe patterns and then make predictions about future levels.

In Activity 1, students explore the Sierpiński carpet. The square shapes in this fractal make it easy for students to calculate areas and perimeters and to investigate how these measures change with each iteration. In Activity 2, students calculate areas of different levels of the Sierpiński triangle. Using isometric dot paper for this activity helps students develop an intuitive sense of the areas based on subdivisions of the initiator triangle. In Activity 3, students explore area patterns in the Pythagorean tree.

The pattern-exploration activities in this chapter informally introduce the idea of a limit, although the word *limit* is not used. Students can see the visual pattern of a series of numbers approaching a value, without the possibility of ever reaching that value.

The remaining activities extend some important mathematical ideas associated with the fractal designs.

Activity 6 introduces another way to create fractals: plotting points within the framework of a polygon. In the Chaos Game, students use a random process to generate the Sierpiński triangle. The game is a reminder of the connection between fractals and chaos theory. Mathematicians and scientists study chaotic events such as earthquakes, hurricanes, and even wars to determine if an underlying pattern emerges over time. The Chaos Game gives players firsthand experience with this phenomenon.

This Chaos Game works well if groups play on triangles copied on overhead transparencies with transparency markers. After each group has plotted 20 or 30 points, lay the transparencies on top of one another to see the Sierpiński triangle pattern. After students have completed Activity 2, introduce them to the calculator version of the game in Activity 7. A calculator can plot hundreds of points in only a few seconds, making the Sierpiński triangle pattern apparent much more quickly. If students enjoy these activities, you might encourage them to research the relationship between chaos and fractals. Technically, the Chaos Game is misnamed—*mathematical chaos* is the production of seemingly random patterns by methods that are not random. Thus, computers can be programmed to create what looks like a mountain range. In the Chaos Game, the process of creating the pattern is random, yet no pattern emerges.

The Sierpiński carpet and Sierpiński triangle are considered *sieves* because parts of the original design are systematically removed or "sifted out" with each iteration. Activity 4 introduces students to a numerical sieve called the sieve of Eratosthenes. This sieve removes the composite numbers from the set of natural numbers, allowing the prime numbers to be identified. In this version of the sieve, students mark the multiples of each prime in a unique way. The markings allow students to look for patterns in the multiples and to identify prime factors and common factors. The sieve of Eratosthenes is a useful tool when students use prime factorization to find least common multiples and greatest common factors. The branch of mathematics that explores patterns in numbers such as the patterns the sieve reveals is called *number theory*.

The Pythagorean tree design is a visual model of the Pythagorean theorem. In Activity 5, students use the shapes in the tree design to explore the theorem.

Activity 8 provides a list of topics for students to explore on the Internet. Caution students that information from Web sites may not be accurate. They will need to use other sources to verify the information.

Answers and Comments

Activity 1

1.

	Area Removed (square units)	Area (square units)	Perimeter (including perimeters of removed squares or units)
Level 0 (initiator)	0	$27 \cdot 27 = 729$	$4 \cdot 27 = 108$
Level 1	$9 \cdot 9 = 81$	$729 - 9 \cdot 9 = 648$	$108 + 4 \cdot 9 = 144$
Level 2	$81 + 8 \cdot 3 \cdot 3 = 153$	$648 - 8 \cdot 3 \cdot 3 = 576$	$144 + 8 \cdot 4 \cdot 3 = 240$
Level 3	$153 + 8 \cdot 8 \cdot 1 \cdot 1 = 217$	$576 - 8 \cdot 8 \cdot 1 \cdot 1 = 512$	$240 + 8 \cdot 8 \cdot 4 \cdot 1 = 496$

2. Possible answer: The area is decreasing while the perimeter is increasing. This is because removing a square decreases the area but adds interior perimeter.
3. $\frac{\text{area of Level 1}}{\text{area of Level 0}} = \frac{8}{9} = 0.\overline{8}$

   $\frac{\text{area of Level 2}}{\text{area of Level 1}} = \frac{8}{9} = 0.\overline{8}$

   $\frac{\text{area of Level 3}}{\text{area of Level 2}} = \frac{8}{9} = 0.\overline{8}$

   The ratios are the same because $\frac{1}{9}$ of the area is removed with each iteration.

Activity 2

1. Each iteration bisects the sides. Because dividing a positive integer power of 2 by 2 results in a whole number, starting with side lengths that are a power of 2 will give whole-number side lengths for the first few levels. If you start with sides of length 2^n, then you can draw n levels of the fractal without having to use fractional side lengths.
2. a.

	Area in Fraction Form (sq. units)	Area in Decimal Form (sq. units)	Total Area Removed (sq. units)
Level 0	1	1	0
Level 1	$\frac{3}{4}$	0.75	$\frac{1}{4}$
Level 2	$\frac{3}{4} \cdot \frac{3}{4} = \frac{9}{16}$	0.5625	$\frac{1}{4} + \frac{1}{4} \cdot \frac{3}{4} = \frac{7}{16}$
Level 3	$\frac{3}{4} \cdot \frac{9}{16} = \frac{27}{64}$	0.421875	$\frac{7}{16} + \frac{1}{4} \cdot \frac{3}{4} \cdot \frac{3}{4} = \frac{37}{64}$
Level 4	$\frac{3}{4} \cdot \frac{27}{64} = \frac{81}{256}$	0.31640625	$\frac{37}{64} + \frac{1}{4} \cdot \frac{3}{4} \cdot \frac{3}{4} \cdot \frac{3}{4} = \frac{175}{256}$

   b. The areas get smaller. The amount removed each time decreases.

   c. Possible answers: The numerators are powers of 3, and the denominators are powers of 4 (and therefore also powers of 2). This is because each unshaded triangle in one level is replaced by three unshaded triangles in the next level, and each of these new triangles has $\frac{1}{4}$ the area of the new triangles in the previous level.

The decimal fraction answers all end in 5 because the divisor contains only prime factors of 10. For the first seven levels, the number of decimal places in the areas equals the denominator's power of 2. For example, in Level 2, the denominator is 16, or 2^4, and the decimal equivalent, 0.5625, has four decimal places.

d. $\left(\frac{3}{4}\right)^0; \left(\frac{3}{4}\right)^1; \left(\frac{3}{4}\right)^2; \left(\frac{3}{4}\right)^3; \left(\frac{3}{4}\right)^4$

e.

	Area in Fraction Form (sq. units)	Area in Decimal Form (sq. units)
Level 5	$\frac{243}{1024}$	0.2373046875
Level 6	$\frac{729}{4096}$	0.177978515625

The numbers are getting closer to 0. More and more area is being removed, so after some area has been removed infinitely many times, none will remain.

Activity 3

	Level 0 (initiator)	Level 1	Level 2	Level 3	Level 4	Level 5
Total number of squares	1	3	7	15	31	63
Area of each new square	256	128	64	32	16	8
Total area of all squares	256	512	768	1024	1280	1536
Total number of triangles	1	3	7	15	31	63
Area of each new triangle	64	32	16	8	4	2
Total area of all triangles	64	128	192	256	320	384
Total area of level	320	640	960	1280	1600	1920

1. Possible answer: Finding an area was easiest when the base and height were vertical and horizontal segments because you could count grid units to find the measurements. All triangles were easy because either the legs were horizontal and vertical or the hypotenuse and the altitude to the hypotenuse were horizontal and vertical. The nontilted squares were easy because the sides were horizontal and vertical. The tilted squares were difficult because you could not count grid squares to find the side lengths.

2. The area of each triangle is $\frac{1}{4}$ the area of the square it is constructed on.

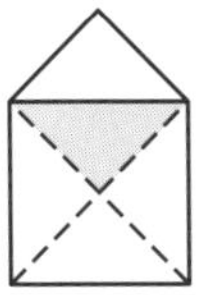

3. Possible answer: The area of each new square is half the area of each square that was new in the previous level. The total area of the squares is always 256 more than the total area of the squares in the previous level.
4. Possible answer: The area of each new triangle is half the area of each triangle that was new in the previous level. The total area of the triangles is always 64 more than the total area of the triangles in the previous level.
5. The total area keeps growing by the original area, 320, with each level. At each level, two smaller squares and two smaller triangles are added to each square and triangle. The smaller squares are each made up of two triangles of the same area as the triangle, which was $\frac{1}{2}$ the square. The area of each smaller triangle is $\frac{1}{2}$ the area of the triangle.

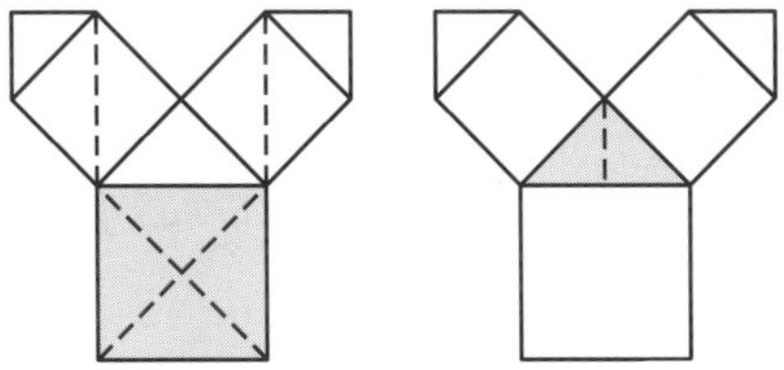

Activity 4

Remind students that in mathematics, the term *sieve* describes a process that removes or "sifts out" areas or numbers. The Sierpiński carpet and Sierpiński triangle are sieves because each iteration removes area from a polygon. In this activity, once students have marked the multiples of all numbers through 30, all unmarked numbers up to 60 are primes. Students can then determine the prime factors of each composite number in the sieve by looking at the marks in its square. For example, the square for 12 includes marks for 2 and 3, so 2 and 3 are the prime factors of 12. Every composite number can be written as the product of prime numbers. For example, $12 = 2 \times 2 \times 3$.

TEACHER NOTES

For students who are having trouble completing the sieve, you might work with them through the multiples of 3, move to the multiples of 4 and see if it is already marked, then work through the marking of multiples of 5.

1. 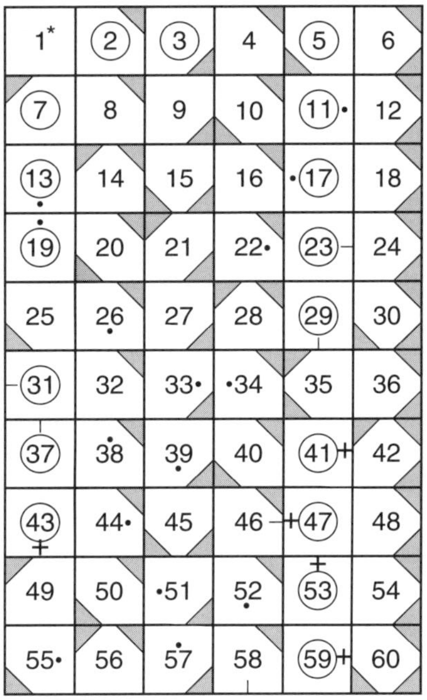

2. 2, 3, 5, 7, 11, 13, 17, 19, 23, 29, 31, 37, 41, 47, 53, 59
3. Multiples of 2 appear in the second, fourth, and sixth columns. Sifting out multiples of 2 removes half the numbers.
4. Multiples of 3 occur only in the third and sixth columns. The multiples of 3 in the sixth column are also multiples of 2 ($2 \times 3 = 6$).

   Multiples of 5 occur in two diagonal patterns: starting with 5 and repeatedly moving down one square and left one square, and then starting with 30 and repeatedly moving down one square and left one square.

   Multiples of 7 occur in two diagonal patterns: starting with 7 and repeatedly moving down one square and right one square, and then starting with 49 and moving down one square and right one square.

   Multiples of 11 occur in a diagonal pattern: starting with 11 and moving down two squares and left one square.
5. Because this sieve is in six columns, the patterns for 3s, 5s, and 7s are based on the relationship of these numbers to the number 6.

6. a. Five columns—the second, fourth, sixth, eighth, and tenth—would contain multiples of 2.
   b. The multiples of 3 would no longer be in vertical columns. Instead, they would form a diagonal pattern, moving down one square and left one square.
   c. The multiples of 5 would be in two vertical columns—the fifth column and the tenth column.

7.

Number	Prime Factors	Product of Primes
48	2, 3	$2 \cdot 2 \cdot 2 \cdot 2 \cdot 3$
64	2	$2 \cdot 2 \cdot 2 \cdot 2 \cdot 2 \cdot 2$
51	3, 17	$3 \cdot 17$
60	2, 3, 5	$2 \cdot 2 \cdot 3 \cdot 5$
57	3, 19	$3 \cdot 19$

8. 3 and 5; 5 and 7; 11 and 13; 17 and 19; 29 and 31; 41 and 43

Activity 5

1. a. Areas will vary depending on the size of the square drawn.
   b. The area of the large square equals the sum of the areas of the two small squares.
2. Areas will vary. Students should find that the area of the large square is equal to the sum of the areas of the two small squares.
3. Conjectures may vary. Possible conjecture: If you draw squares on the sides of an isosceles right triangle, then the area of the largest square (the square on the hypotenuse) is equal to the sum of the areas of the smaller squares.
4. Answers will vary. The Pythagorean theorem appears to apply to isosceles right triangles.
5. Conjectures will be similar to those in 3 but will not include the word *isosceles*.

Activity 6

You might suggest that students lightly number the first few points to help them keep track of where they are. Once they are used to the process, they can stop numbering. If measurements are correct and the first few points are ignored, the pattern of a Sierpiński triangle will result from the Chaos Game.

If the game begins with an isosceles right triangle, the result will have this pattern.

Activity 7

You can download the programs for the TI-73, TI-82, or TI-83 and TI-83 Plus from **www.keymath.com/DA/calculator_programs.** Students can check with you or other students to confirm that the image on the calculator screen is correct. You might ask students to try different shapes and different fractions. Ask them what effect the fraction has on the pattern that is created.

Activity 8

Reports on the research topics can be written, presented orally, or displayed visually.

The Linear Fractal Quilts

A FRACTAL IS THE RESULT OF A PROCESS THAT BEGINS with a geometric figure (called the *initiator*), applies a transformation to the figure to produce Level 1 (called the *generator*), transforms the generator in the same way to produce a new result, transforms that result in the same way, and so on for infinitely many steps.

Fractals are *self-similar.* That is, parts of the fractal look like the entire fractal, only at a smaller scale.

Unlike the fractals in the previous section, which began with polygons, the fractals in this section begin as line segments or are formed by line segments drawn on polygonal grids. As successive transformations are performed, these fractals wind their way around a two-dimensional surface, seeming to fill all or part of a space enclosed by visible or invisible boundaries. For this reason, these fractals are often called space-filled curves. Mathematicians have studied space-filling curves for over a century. When Benoit Mandelbrot coined the term *fractal* in the 1970s, he included these space-filling curves because they are self-similar.

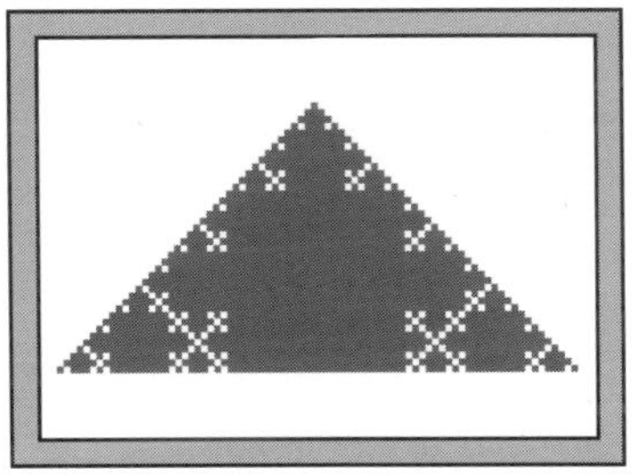

The Koch Coastline Quilt

The Koch Snowflake Quilt

The Hilbert Curve Quilt

The Sierpiński Curve Quilt

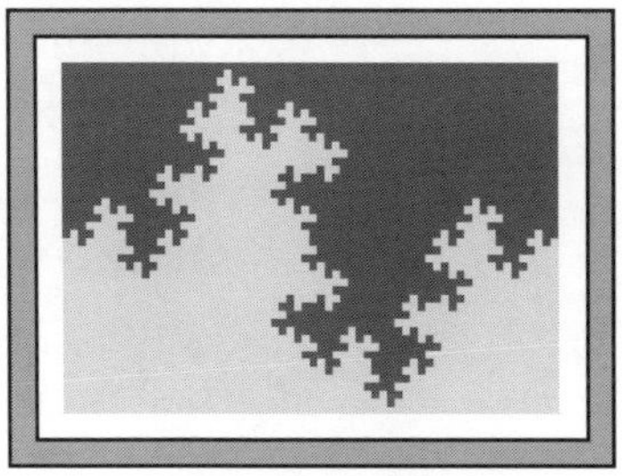

The Three-Halves Quilt

The Koch coastline (pronounced “coke,” but with a kh sound not used in English) is a simple example of a linear fractal.

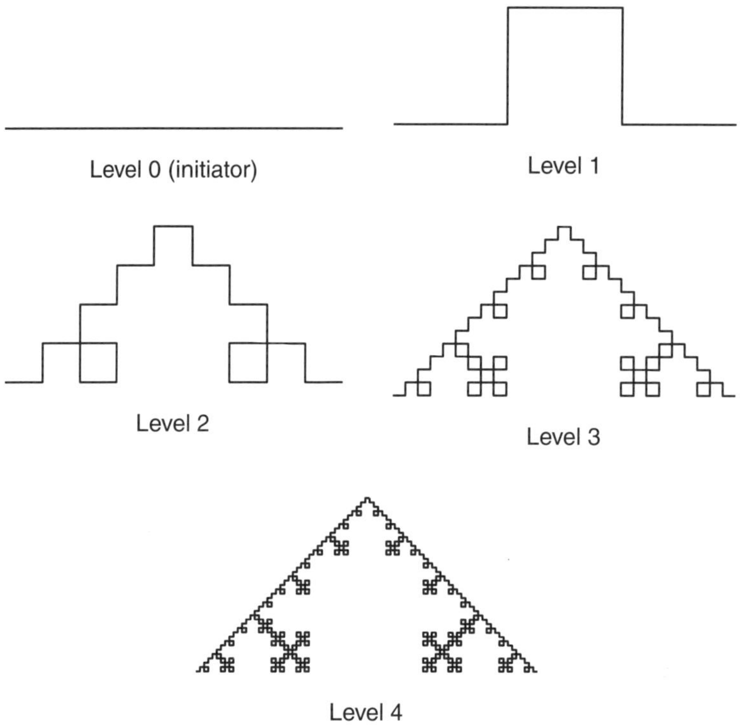

The initiator of this fractal is a line segment. The first iteration replaces the middle third of the segment with three sides of a square, which creates Level 1. The next iteration applies this same transformation to each of the five segments in Level 1, creating Level 2. This process is repeated indefinitely to produce a curve that is longer at each level as smaller pieces are added. As its name suggests, natural coastlines can be analyzed using the tools of fractal geometry.

Fractal designs provide many examples of exponential patterns. For example, in the Koch coastline, the number of segments grows with each level in the exponential pattern 1, 5, 25, 125, and so on. Each level has five times as many segments as the previous level. The number of segments in Level n is 5^n.

Exponential patterns can also be found in the lengths of the segments and in the perimeter, or total length, of the curve. The length of each new segment is $\frac{1}{3}$ the length of a segment in the previous level. So if the length of the initiator segment is 1, then the length of each segment at Level n is $\left(\frac{1}{3}\right)^n$. The total length of the curve at each level is $\frac{5}{3}$ times the length at the previous level. So, the total length of the curve at Level n is $\left(\frac{5}{3}\right)^n$.

The three-halves curve, like the Koch coastline, begins with an initial line segment. To create successive levels of the curve, each segment is divided into four congruent segments, and the two middle segments are each replaced by three sides of a square—one square on each side of the segment.

Level 0 (initiator)

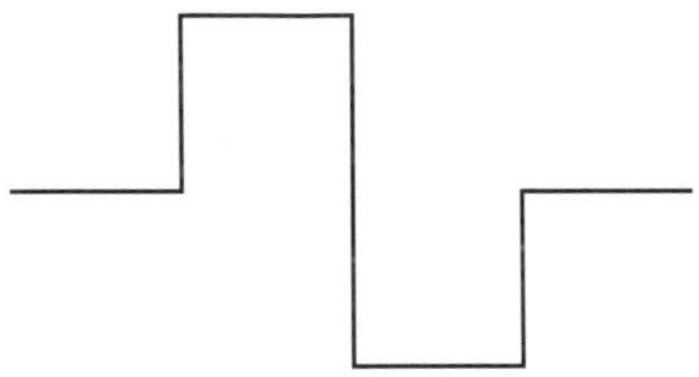

Level 1 (generator)

Level 2

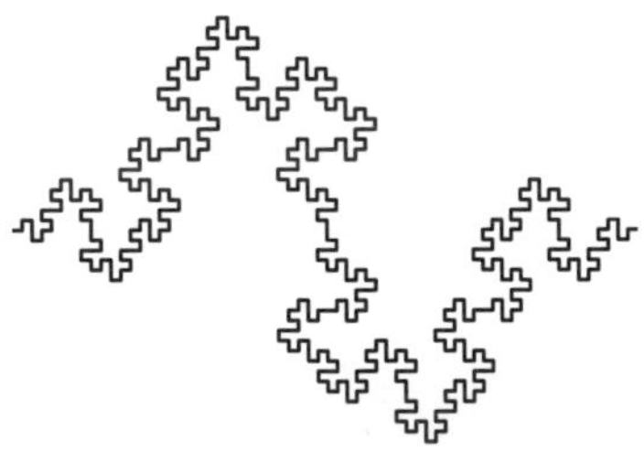

Level 3

The three-halves curve is named for its dimension, which after infinitely many iterations fills the space between that of a line segment with one dimension (length) and the region inside a closed figure with two dimensions (length and width).

When mathematicians studied fractals and their meandering patterns, they sensed that these figures had characteristics of both one- and two-dimensional figures. So they developed methods to assign *fractional* dimensions to fractals, such as using the logarithm of the *scale factor* and the *size factor*. (See Activity 6 to learn more about dimensions of fractals.)

The scale factor is determined by counting the number of segments the initiator segment is divided into. The size factor is the number of new segments that replace the initiator segment. The formula for calculating a fractal's dimension is

$$\text{dimension} = \frac{\log_x(\text{size factor})}{\log_x(\text{scale factor})}$$

For the three-halves curve, the scale factor is 4 because the initiator segment is divided into four congruent segments. The size factor is 8 because the initiator segment is replaced by eight new segments. To calculate the dimension $\frac{\log 8}{\log 4}$, remember that logarithms are exponents, so choose a convenient base of 2. The exponent of 2 that would equal 8 is 3 ($2^3 = 8$), and the exponent of 2 that produces 4 is 2 ($2^2 = 4$). Thus, the dimension is $\frac{\log_2 8}{\log_2 4} = \frac{3}{2}$ (three-halves).

The Koch snowflake consists of three identical fractal designs, one on each side of an equilateral triangle. However, the Koch snowflake is not itself a fractal; there are no smaller parts of the snowflake that look like the whole design. The Koch snowflake is similar to the Koch coastline but differs in that the removed segment is replaced with two sides of an equilateral triangle. This design has been given the name Koch snowflake because it resembles a snowflake as it advances in level; the name is another reminder of the many fractal patterns in nature.

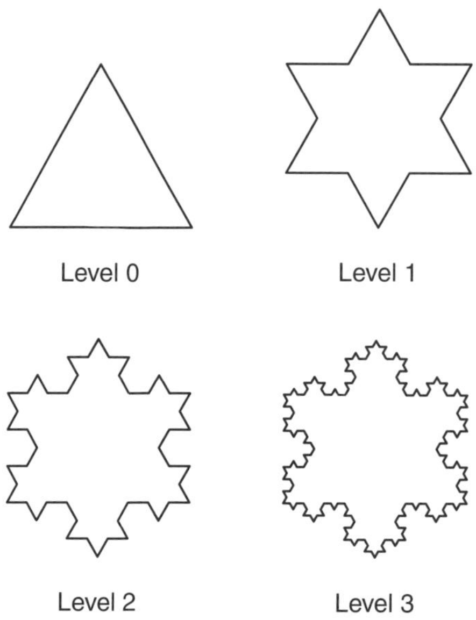

One way to draw the Sierpiński curve and the Hilbert curve is not to build on an initiator line segment, but instead to draw line segments inside a polygon.

For each level of the Sierpiński curve, a square is divided into isosceles right triangles. To create each level of the fractal, start with a new square and divide it into triangles by drawing diagonals and horizontal and vertical segments. Line segments drawn across the segments in the foundation form square and octagon shapes.

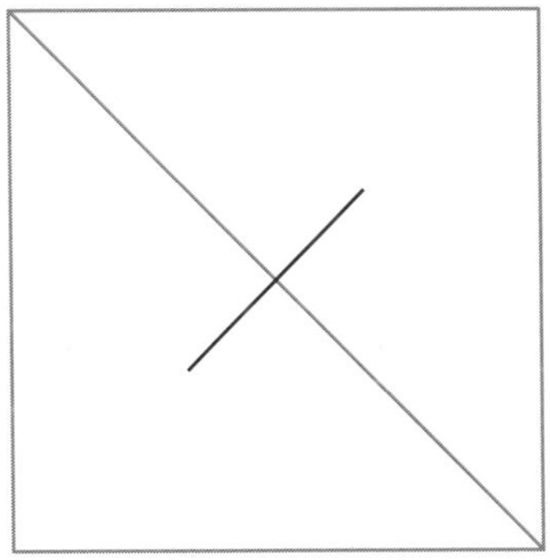

Level 0

Level 1

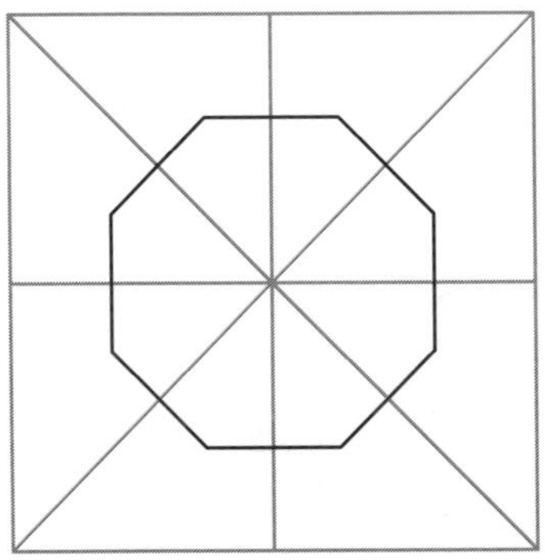

Level 2

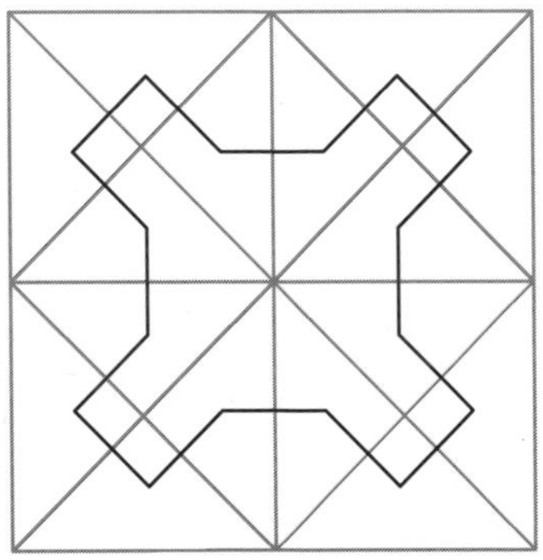

Level 3

The foundation for each level of the Hilbert curve is a square subdivided into smaller squares. The curve follows a path that passes through the four quadrants of a small square before moving onto the next square.

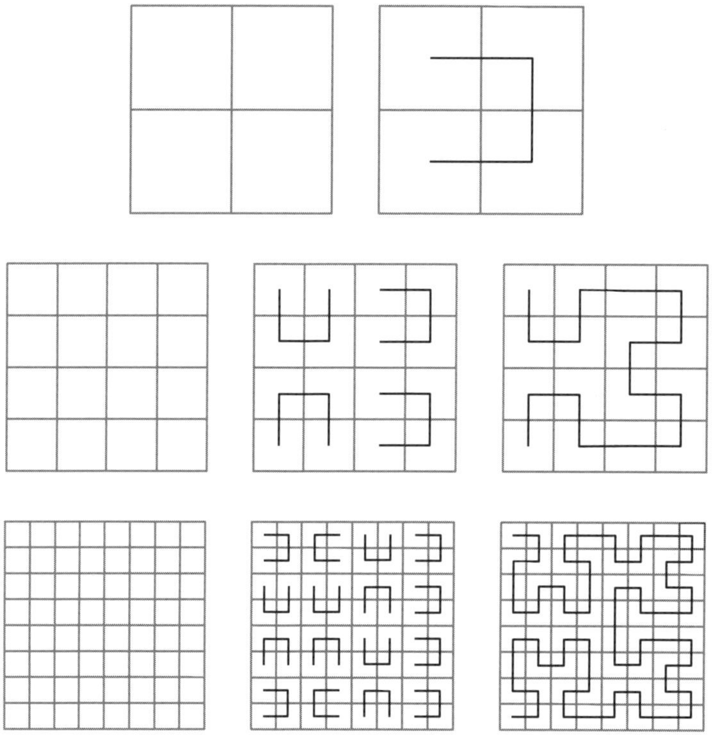

The Koch Coastline Quilt

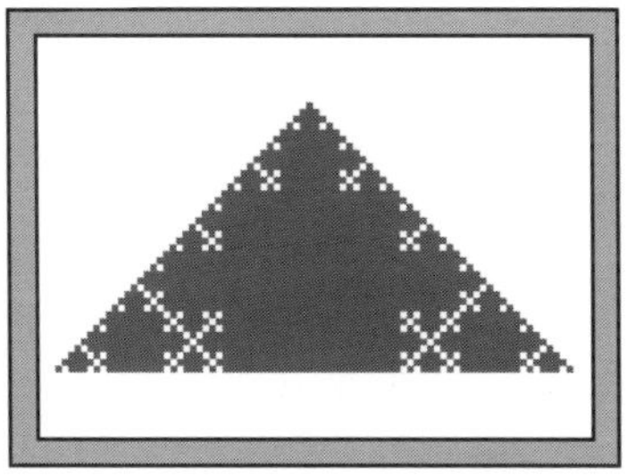

The Koch coastline fractal was created by the Swedish mathematician Helge von Koch (1870–1924). The initiator is a line segment. At each iteration, each segment is divided into three congruent segments; then a square is built on the middle segment, and the middle segment is removed.

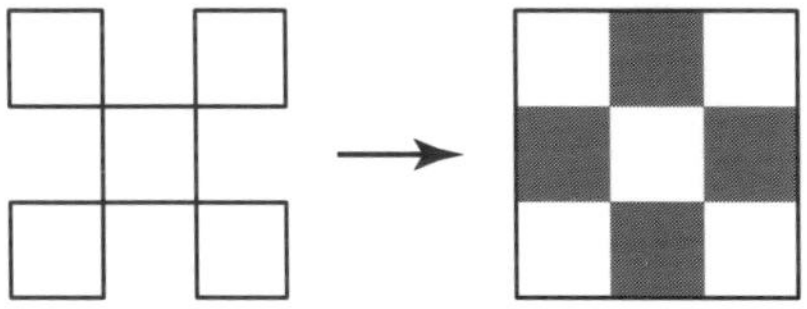

At Level 2 of the fractal, the endpoints of segments meet, creating squares. Starting in Level 3, three-by-three square patterns emerge. When the three-sided "squares" in these patterns are shaded, the checkerboard patterns seen in the quilt are formed. Quilters refer to these patterns as nine-patches.

The Koch coastline quilt shows Level 4 of the fractal design. Red fabric was used for the shaded squares, and white was used for the unshaded squares and the background fabric.

Designing Your Quilt

Follow these steps to create a Koch coastline design. Because creating the design requires erasing segments, be sure to use a pencil.

Step 1 Draw a segment with length 81 units near the bottom of a sheet of rectangular dot paper or graph paper. (Or tape several sheets together.)

Step 2 Divide the segment into three congruent segments.

Step 3 Draw a square with one side on the middle segment, and then erase the middle segment.

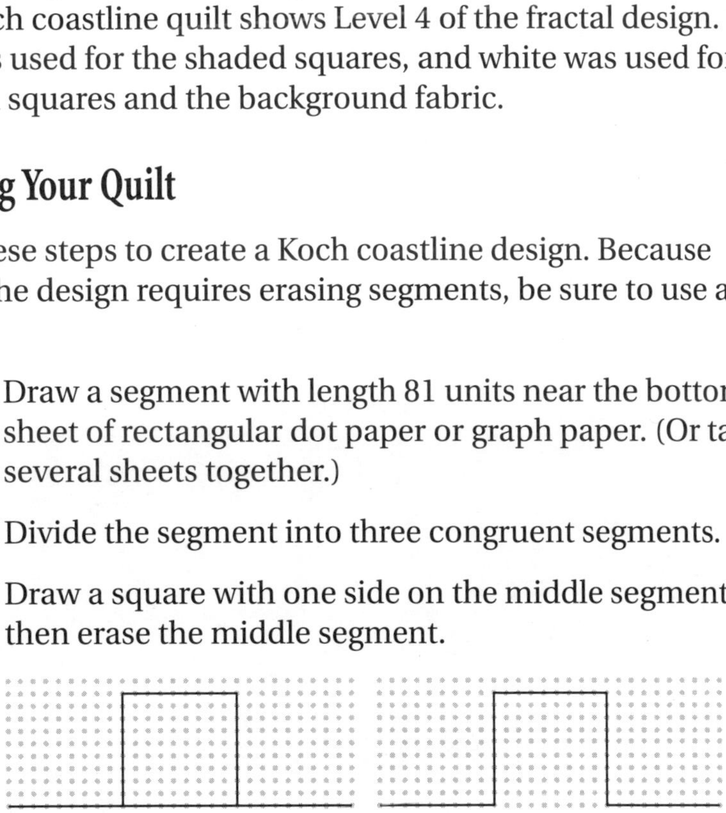

Level 1

Step 4 You have just created Level 1. This is the generator that is repeated to create each level of the design. To create Level 2, repeat Steps 2 and 3 for each of the five segments in Level 1.

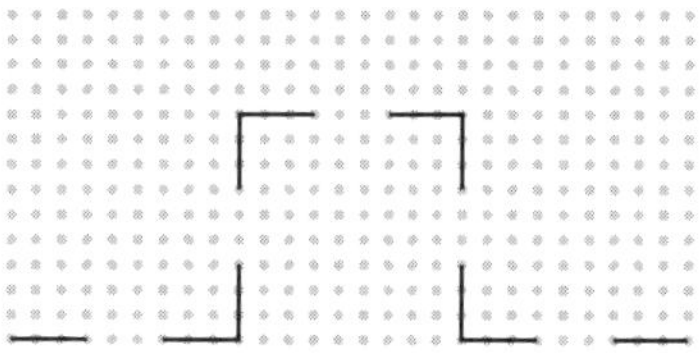

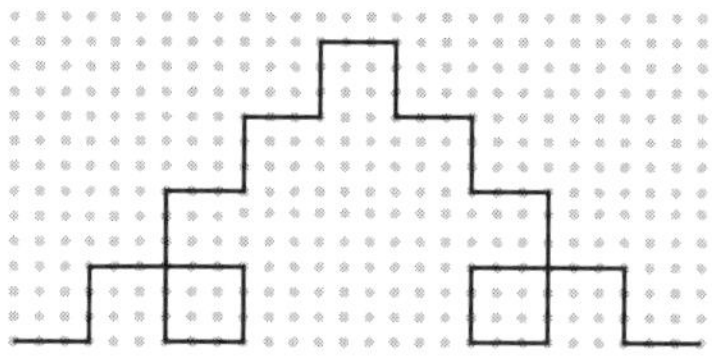

Level 2

Step 5 To draw each succeeding level, follow Steps 2 and 3 for each segment in the previous level.

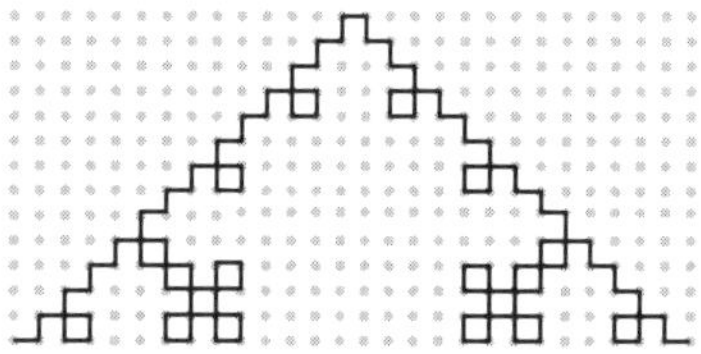

Level 3

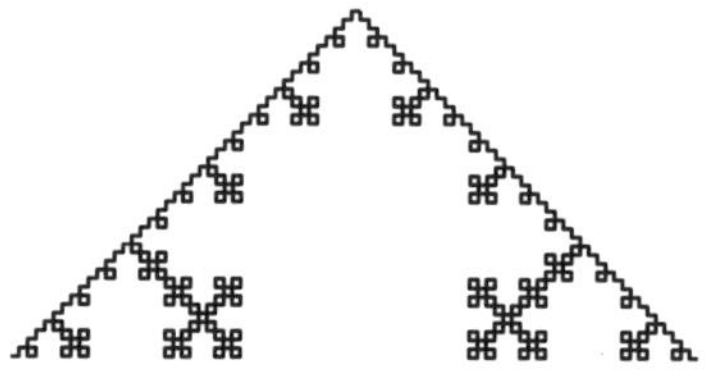

Level 4

Note: To figure out which side of each segment to draw the square on, imagine each level as a continuous path from left to right. Add the square so that after you erase the middle segment, the path takes a left turn, then a right turn, then a right turn, then a left turn.

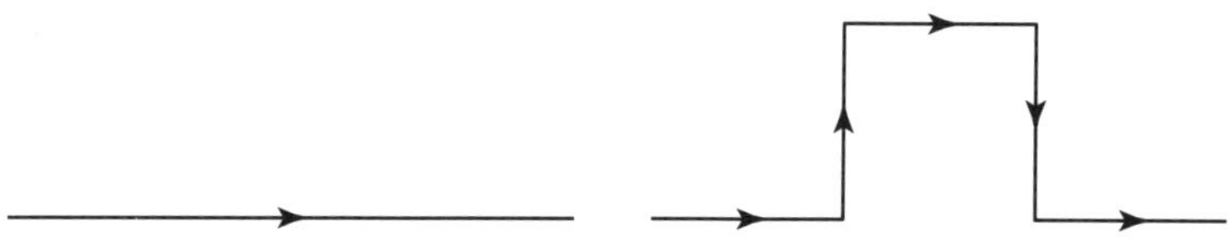

Step 6 Color your design.

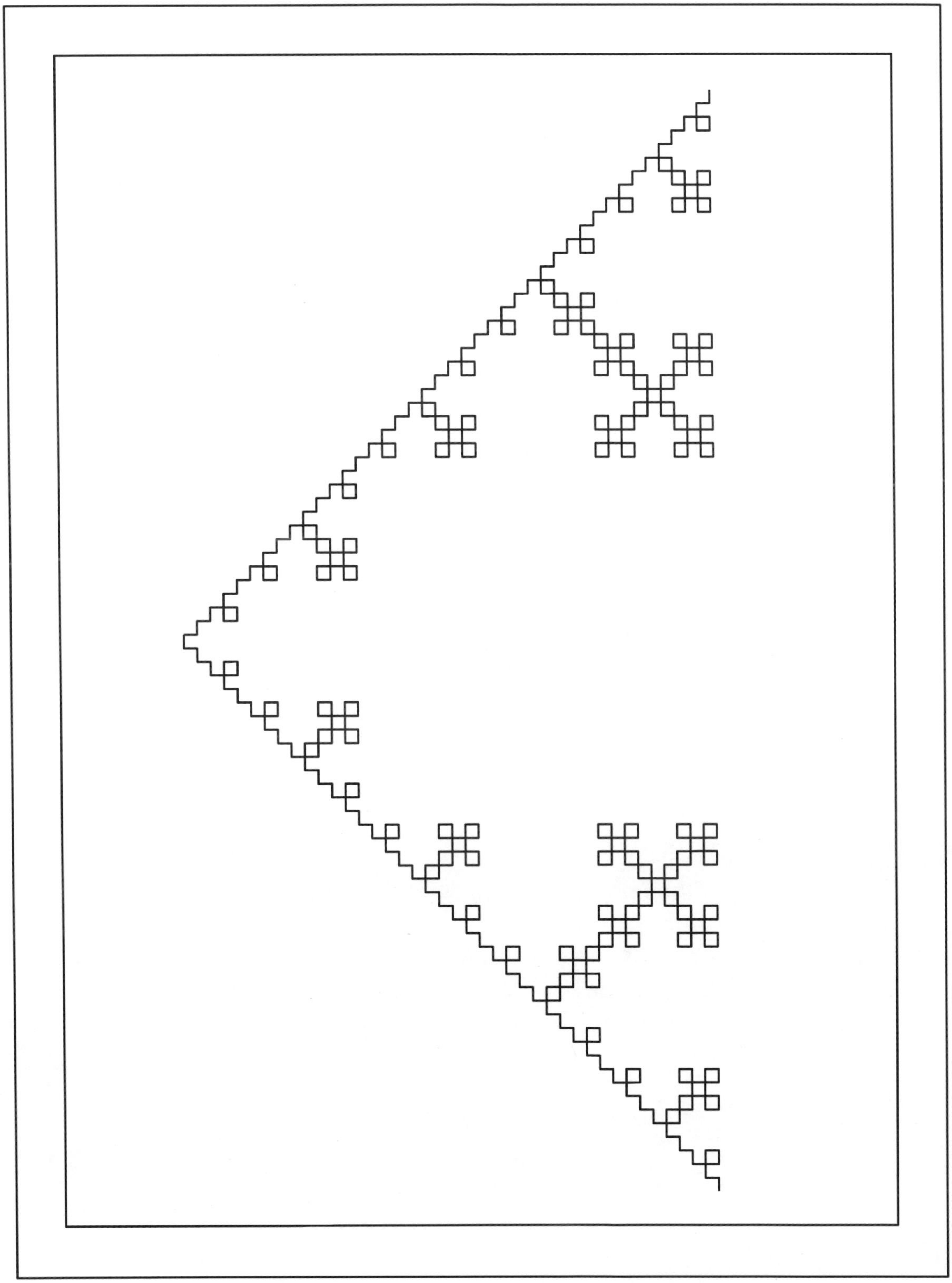

Making the Koch Coastline Quilt

The quilt pictured shows Level 4 of the Koch coastline fractal, in which the side length of the smallest square is $\frac{1}{2}$ inch. The finished size of the quilt is 23 by 42 inches. The instructions that follow are for a Level 3 quilt in which the side length of the smallest square is 1 inch. The width of the quilt will be 27 inches.

Materials

$1\frac{1}{4}$ yards of a light-colored background fabric

$\frac{1}{2}$ yard of a dark-colored fabric

1 yard of backing fabric

$\frac{1}{2}$ yard of binding fabric

$\frac{3}{4}$ yard of batting

Instructions

This table lists the pieces needed for the quilt. The cut sizes allow for a $\frac{1}{4}$-inch seam allowance.

Cut Size	Finished Size	Number of Dark Pieces	Number of Light Pieces
1.5″ × 1.5″	1″ × 1″	25	18
3.5″ × 3.5″	3″ × 3″	5	0
9.5″ × 9.5″	9″ × 9″	1	0

Cut a 36-by-44-inch piece of the background fabric. Using a light pencil line, draw a horizontal line across the background fabric approximately 10 inches from the bottom. Lay the large dark-colored square on the background fabric and position the other pieces to re-create Level 3 of the Koch coastline.

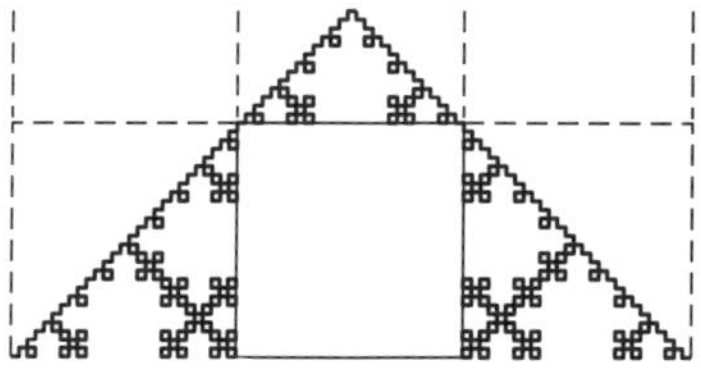

Quilt blocks

Visualize the design as having a left, right, middle, and top section. Using a $\frac{1}{4}$-inch seam allowance, sew the squares of each section to each other where the edges are adjacent. Join the sections where the edges are adjacent.

Baste unstitched edges under with a $\frac{1}{4}$-inch seam allowance. Appliqué the design to the background fabric. (See *Mastering Machine Appliqué* by Hargrave for specific instructions.)

Baste the quilt to the batting and backing, and then quilt the layers together to duplicate the design of the fractal, alternating nine-patches with solid squares.

Finish the quilt by binding the perimeter. Add a rod pocket to hang the quilt if desired.

The Three-Halves Curve Quilt

The initiator of the three-halves curve is a line segment. At each iteration, the segment is divided into four equal parts, squares are drawn on the two middle segments—one on either side of the original segment—and then the two middle segments are removed.

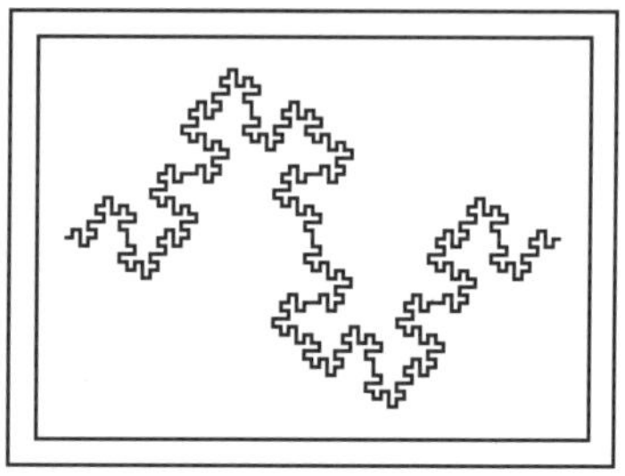

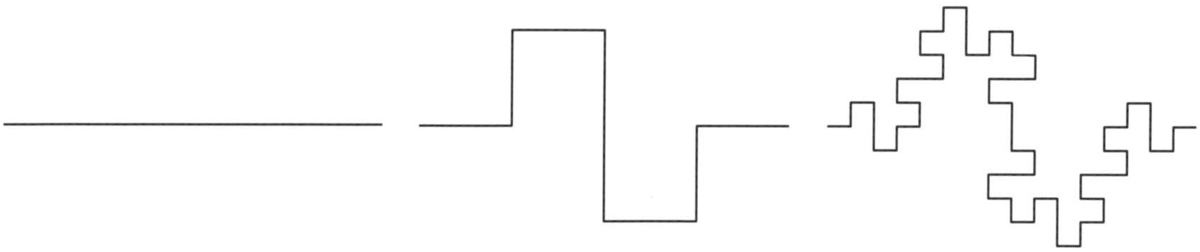

Level 1 Level 2

The three-halves curve is also known as *Minkowski's curve* or *Minkowski's sausage,* named by Benoit Mandelbrot to honor Hermann Minkowski (1864–1909), a fellow geometer.

The three-halves curve quilt shows Level 3 of the fractal. Contrasting light and dark fabrics were chosen for this quilt to accentuate the intricate design of the fractal. The dark fabric has a black background with random streaks of tan. The light fabric is beige with random streaks of brown. These two fabrics, when placed together, create a dramatic effect.

Designing Your Quilt

To draw the three-halves curve to Level 3, you will need rectangular dot paper or graph paper measuring at least 64 by 42 units. Tape two or more sheets together if necessary. Because creating the design requires erasing segments, be sure to use a pencil.

Step 1 Draw a line segment 64 units long, centered vertically, so there is the same amount of space above and below.

Step 2 Divide the segment into four equal segments.

Step 3 Draw squares on the two middle segments so that one square is on one side of the original segment and one square is on the other side. Then erase the two middle segments.

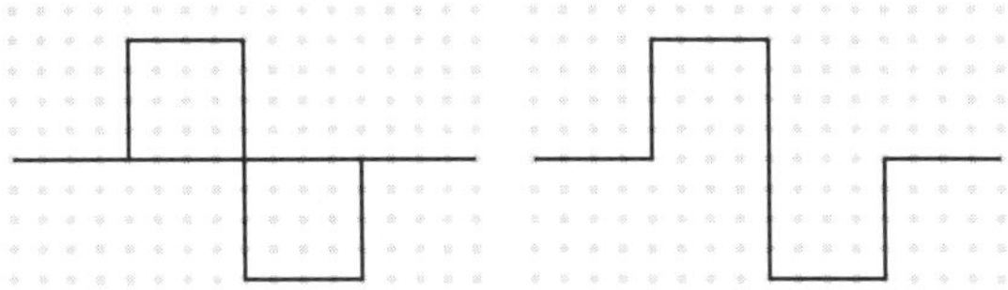

Level 1

Step 4 You have just created Level 1 of the design. This is the generator that is repeated to create each level of the design. To create Level 2, repeat Steps 2 and 3 for each of the eight segments in Level 1. (Think of the long vertical segment as two shorter segments.)

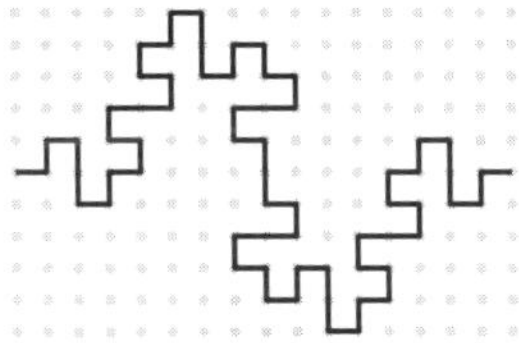

Level 2

Step 5 To draw each succeeding level, follow Steps 2 and 3 for each segment in the previous level.

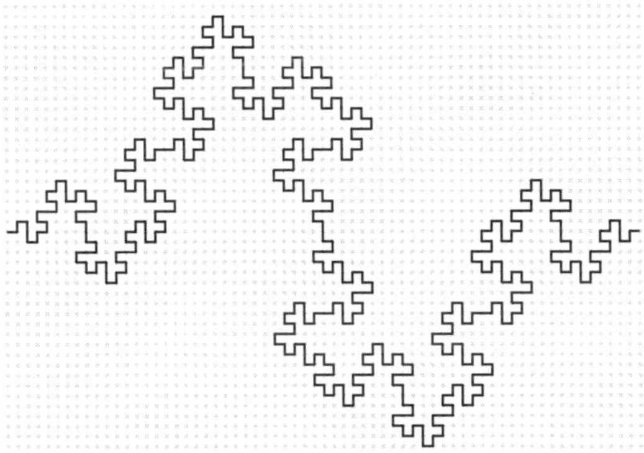

Level 3

Note: To figure out which side of each segment to draw the squares on, imagine each level as a continuous path from left to right. Add the squares so that after you erase the middle segments, the path turns left, then right, then right, then left, then left, then right.

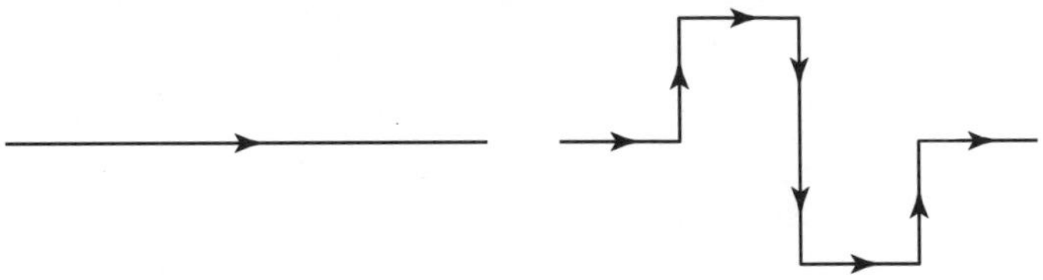

Step 6 Color your design.

Making the Three-Halves Curve Quilt

The finished three-halves curve quilt is 28 by 33 inches. The quilt was made with a layering technique used by David Walker, a contemporary quilter. Walker layers many fabrics, right side up, and stitches them together with design lines that divide the surface into many sections. He then cuts away layers of fabric in each section to create a patchwork effect. He completes his designs with embellishments such as glittering threads, beads, and buttons.

Materials

Large sheets of graph paper and tracing paper

1 yard each of two fabrics for the quilt top

1 yard of backing fabric

1 yard of batting

$\frac{1}{2}$ yard of fabric for binding

Thread to match dark fabric

Instructions

Follow the instructions in the "Designing Your Quilt" section on page 54 to create a full-size template of Level 3 on graph paper. Start with a segment 32 inches long. (The final design will be made up of $\frac{1}{2}$-inch segments.) Trace the template onto tracing paper.

Layer the two fabrics right side up with the lighter fabric on top. Press the fabrics and pin together in a few places around the edges to help the layers stay in position. Pin the tracing-paper template to the top of the layered fabrics. Adjust the template so it is centered on the fabric and the horizontal and vertical segments align with the grain lines of the fabric. Place pins as close to the fractal lines as possible while leaving room for the sewing machine to maneuver around them. (The tracing paper will act as a stabilizer as well as a template. If your machine requires more stabilizer for satin stitching, place tear-away stabilizer under the layered fabric.)

Choose a bold thread color for the satin stitching so that the fractal design will dominate the quilt top. Set the machine for satin stitch with a width of $\frac{1}{8}$ to $\frac{3}{16}$ inch. You will be stitching half-inch segments, turning the fabric 90 degrees after each segment, so test different widths to find the one that works best for your machine.

Begin at one end and satin stitch the fractal design on top of the tracing paper. (It is a good idea to practice satin stitching short perpendicular segments before you work on the quilt top.) When you have finished stitching all the segments, tear away the tracing paper.

Using small scissors or appliqué scissors, carefully cut away the top layer of fabric along one side of the satin stitching line. You may need to cut once and then return to snip any threads that have unraveled.

At this point, the satin stitching may be refreshed by sewing over the original stitches with a slightly wider stitch. This second stitching will cover any loose threads or paper.

Layer the quilt top with batting and backing. Quilt any design you feel is appropriate for the fabric. Then trim the fabric, leaving a seam allowance on the left and right sides of the fractal and enough fabric above and below the fractal to create a visually appealing design. Add binding to the perimeter of the quilt to complete it. Add a rod pocket to hang the quilt if desired.

The Koch Snowflake Quilt

The Koch snowflake consists of three identical linear fractal designs, one on each side of an equilateral triangle. At each iteration, each segment is divided into three congruent segments, an equilateral triangle is built on the middle segment, and then the middle segment is removed.

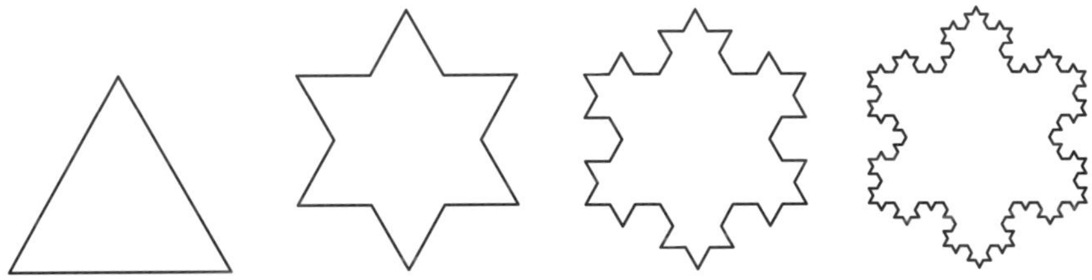

Level 0 Level 1 Level 2 Level 3

The snowflake design becomes more complex as more iterations are performed.

The Koch snowflake quilt was made with a black background, and gold soutache trim was attached to show several levels on one quilt. A white background with blue soutaching is also attractive. The number of levels of the fractal can vary. This quilt shows Level 0 (the initiator level) up to Level 4. Fewer levels can be used.

Designing Your Quilt

Follow these steps to create a Koch snowflake design. Because creating the design requires erasing segments, be sure to use a pencil.

Step 1 Draw an equilateral triangle in the center of a sheet of isometric dot paper. Because the design is formed by trisecting segments, use a side length that is a power of 3. If you start with side lengths of 3^3, or 27, units, you can draw the design to Level 3 without using fractional segment lengths.

Step 2 Divide each segment into three congruent parts.

Step 3 Draw an equilateral triangle on the middle third of each segment so the triangle is on the *exterior* of the figure, and then erase the middle segment.

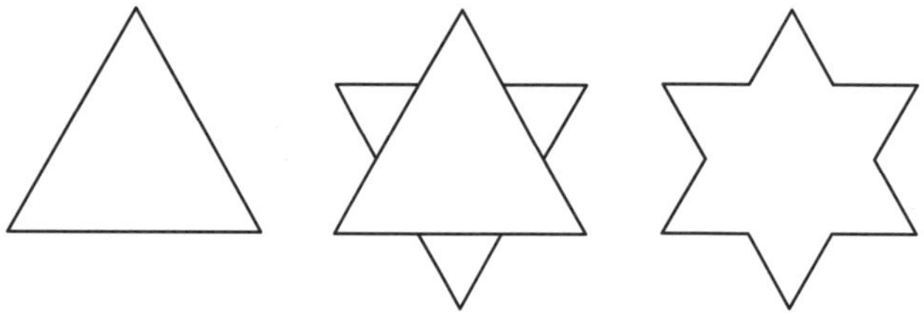

Level 0 Level 1

Step 4 To create each succeeding level, repeat Steps 2 and 3 for each new segment from the previous level.

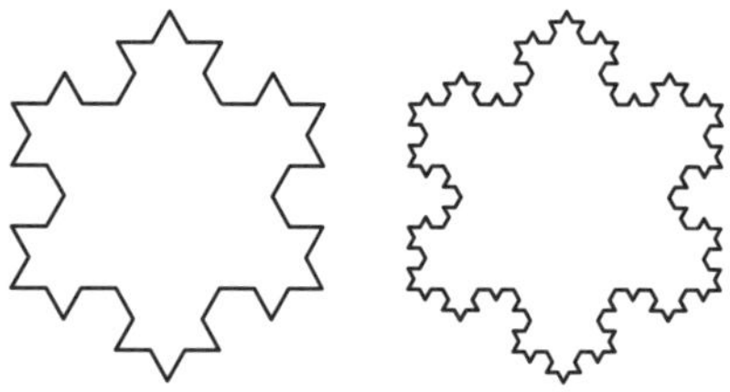

Level 2 Level 3

Making the Koch Snowflake Quilt

The finished Koch snowflake quilt is 30 by 33 inches.

Materials

1 yard of fabric for the front of the quilt

1 yard of fabric for the back of the quilt

40 feet of gold cording

4 yards of gold bias tape made from a stretchable material

Instructions

To create this quilt, gold cording is used to outline the fractal design. The cording is attached by hand.

The quilt shows Levels 0 through 4 of the fractal design. Draw the five levels on graph paper with a common center or freezer paper. The center triangle should measure $3\frac{1}{2}$ inches on each side. For the Level 1 figure, start with a triangle measuring 7 inches on a side. The initiator triangles for Levels 2 through 4 should have sides measuring $10\frac{1}{2}$ inches, $16\frac{1}{2}$ inches, and $24\frac{1}{2}$ inches, respectively.

After you have drawn the design, transfer it to the front of the quilt. If you drew the design on the dull side of a sheet of freezer paper, iron it onto the surface of the quilt, machine-stitch on the design line, and tear away the paper.

If you did not use freezer paper, you can transfer the design using a tracing pen or pierce the paper at key points and mark those points to be connected once the paper is removed. Cover the sewn or traced design with the heavier cording, sewing it on by hand. It may help to use a stabilizer on the back of the surface fabric to keep the fractal lines straight.

After the entire design is outlined with cording on the front of the fabric, add batting and backing, and quilt around the cording design. Add any embellishments you desire. Bind the quilt edges with the gold bias tape. Add a rod pocket for hanging if desired.

The Sierpiński Curve Quilt

The Sierpiński curve is a space-filling curve whose lines form parts of squares and octagons in alternating levels. Sierpiński's curve is formed within a square, and as this fractal proceeds to higher levels, its segments appear to be filling the space within boundaries.

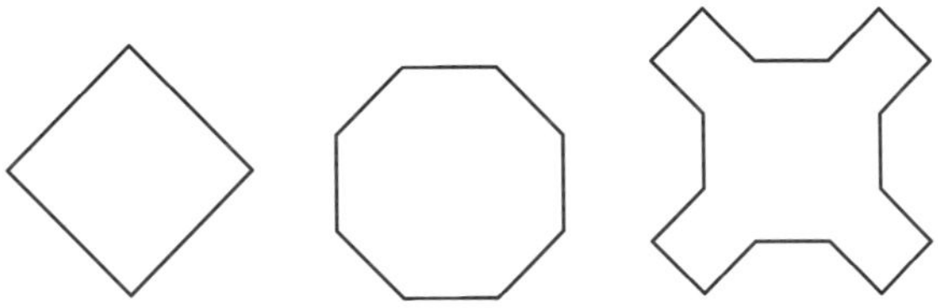

Level 1 Level 2 Level 3

Sierpiński's curve is different from the previous fractals in that each level is created upon a newly constructed foundation. The foundation consists of congruent isosceles right triangles, which form squares in groups of four or eight. The sides of a Sierpiński curve are not formed upon the sides of the previous level; instead, its sides enclose a new shape at each level. The first level is a square; the second level is an octagon. The third level is a shape dominated by protruding squares. The fourth level is dominated by protruding octagons. This alternating pattern continues as higher levels are created.

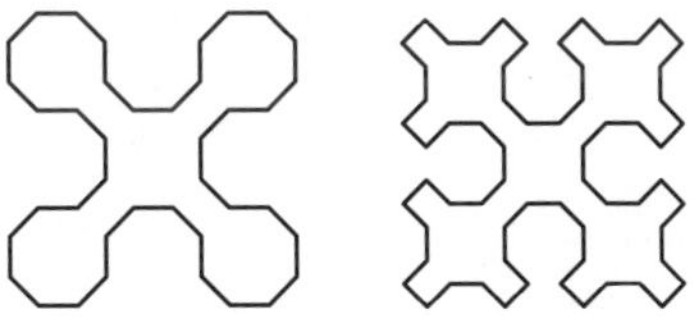

Level 4 Level 5

The quilt is a Level 6 Sierpiński's curve with Level 2 curves in the corners. As this fractal advances, an elegant, symmetric design appears. A variety of colorful calico prints were used for this quilt. The triangle patterns were arranged in a dark/light design as a foundation for the intricate fractal design.

Designing Your Quilt

The Sierpiński curve fractal is a difficult fractal to create because it is formed on a new foundation at each level, and the previous levels are not as much of a guide as they are in the other quilts in this section.

To create the different levels of the Sierpiński curve, you will need several copies of a large square drawn on rectangular dot paper or graph paper. Drawing the levels involves bisecting segments, so use a side length that is a power of 2, such as 32 or 64.

The initiator level is a square with a diagonal as the foundation on which a segment is drawn.

Start with one copy of the square and follow these steps to form Level 1.

Step 1 Draw both diagonals to form four congruent isosceles right triangles. The resulting figure is the foundation for Level 1.

Step 2 Draw short, congruent, perpendicular line segments across the short sides of the triangles. Each segment should be drawn approximately $\frac{1}{3}$ of the distance from the center of the square to the exterior vertex. Use the figure as a guide in drawing the first level.

Level 1

Start with a second copy of the original square and follow these steps to create Level 2.

Step 3 As for Level 1, draw the two diagonals in the square.

Step 4 Now add one horizontal and one vertical segment, each connecting the midpoints of the sides of the square. The square should now be divided into eight congruent isosceles right triangles.

Step 5 Form a regular octagon by drawing congruent line segments across the eight short interior segments that extend from the center of the square. These segments will be approximately $\frac{1}{3}$ the distance from the center of the square to the vertices. Sketch these segments lightly in pencil until your octagon seems to be regular.

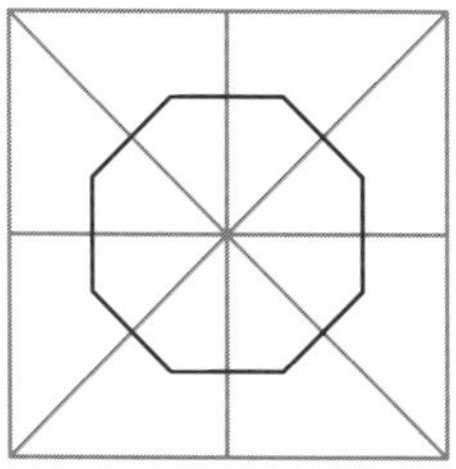
Level 2

To create Level 3, start with a new copy of the original square.

Step 6 Repeat Steps 3 and 4.

Step 7 Each of the squares inside the larger squares has one diagonal drawn. Now draw the other diagonal in each of the four squares. The square now contains 16 congruent isosceles right triangles.

Step 8 In each of the four smaller squares, follow Step 2 to create a square.

Step 9 Erase the side of each small tilted square that is closest to the center. Then connect these squares with horizontal and vertical segments as shown.

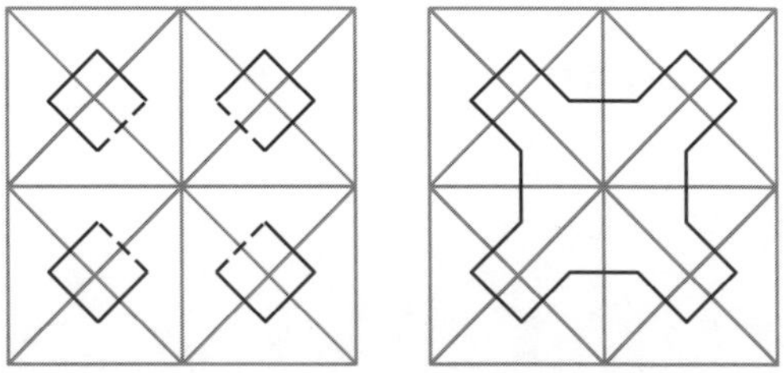

Notice that for the Level 1 foundation, diagonals of a square were drawn. For the Level 2 foundation, horizontal and vertical segments were added. For the Level 3 foundation, diagonals were again drawn. This alternating pattern of drawing diagonals and drawing horizontal and vertical segments repeats for each level.

Start with a fourth copy of the original square and follow these steps to create Level 4.

Step 10 Follow Steps 6 and 7.

Step 11 Add horizontal and vertical segments connecting the midpoints of the sides of each of the four smaller squares.

Step 12 Notice that the figure now has four smaller copies of the foundation used to create Level 2. In each copy of the foundation, draw an octagon as described in Step 5.

Step 13 Erase the side of each octagon closest to the center of the figure, and connect the octagons with horizontal and vertical segments.

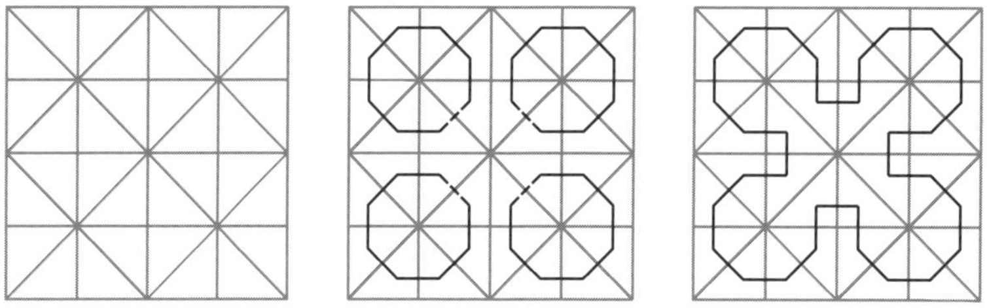

Level 4

Continue to follow the same pattern to create higher levels.

Level 5

Step 14 Color your design.

WORKSHEET 8

Making the Sierpiński Curve Quilt

The quilt has Level 6 of the Sierpiński curve in the center and Level 2 Sierpiński curves at the corners. This quilt measures 18 by 18 inches. The center design of the quilt was made from 128 congruent isosceles right triangles whose legs measured 1 inch after the seams were sewn. The instructions provided are for a quilt at Level 4, made with 32 isosceles triangles whose legs measure $2\frac{1}{2}$ inches after sewing. It will have a finished size of about 10 by 10 inches. To make a larger quilt, use larger triangles or use 64 triangles to make Level 5, or 128 triangles to make Level 6.

Materials

Gridded plastic or heavy paper to create a triangle template

Tracing paper

Enough scraps of dark, medium, and light cotton fabric to create 32 isosceles right triangles, using the template

Dark thread

$\frac{1}{2}$ yard of batting

$\frac{1}{2}$ yard of backing fabric

Binding fabric

Instructions

Create a template for the triangles, using a sheet of gridded plastic or cardboard. Draw an isosceles right triangle whose legs measure $2\frac{1}{2}$ inches. Then add $\frac{1}{4}$-inch seam allowance to all three sides of the triangle.

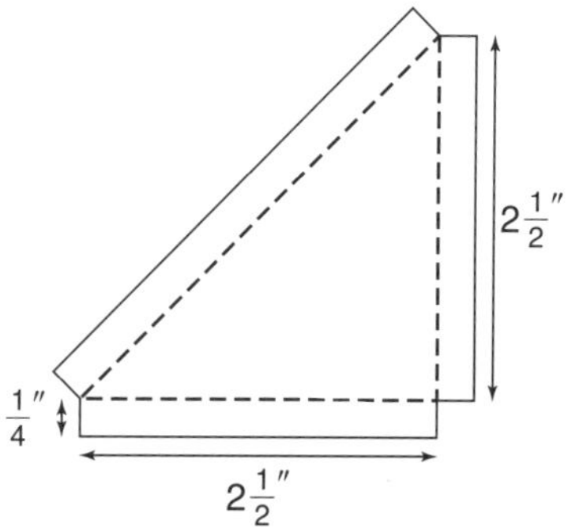

Cut out the template and trace it on the back of 32 pieces of fabric. To create a design with the triangles, cut 16 pieces from the dark or medium fabric and 16 from a light fabric. Before sewing the triangles together, create a design for your foundation on paper. Stitch the triangles together along the hypotenuses, using a $\frac{1}{4}$-inch seam allowance to form 16 squares. Because this seam involves sewing two bias edges together, be careful not to stretch the fabric. Press this seam open. Arrange the 16 squares in the design you have created and stitch them together, row by row. Press the seams open.

On a large sheet of graph paper, re-create the fabric foundation you just made. The squares and triangles on the paper *must* be the same size as those in your fabric design. Using the "Designing Your Quilt" instructions on page 66, create the Sierpiński curve at Level 4. Draw the octagons one at a time, and if one seems to fit perfectly, trace and transfer it to the other octagon locations. When you are satisfied with your design, carefully trace it and the foundation lines onto tracing paper, using a dark pen for the fractal design and a pencil for the foundation.

Pin the traced design onto the fabric, carefully matching the triangles on the paper foundation with the triangles on the fabric. Using a dark thread, satin stitch over the tracing paper lines, carefully turning as the design turns. When you are finished, tear away the tracing paper, removing any small bits of paper. Then restitch the curve with a slightly wider stitch, covering the original stitching and any small bits of paper that cannot be removed.

Add a border, then layer the quilt top over slightly larger batting and backing fabric. Add a quilting pattern if desired. The pictured wall hanging is machine-quilted along both sides of the satin stitching to elevate the stitching. A zipper-foot may be necessary for this step to ensure the stitching is close to the satin stitching. Add a binding and a rod pocket for hanging.

The Hilbert Curve Quilt

The Hilbert curve is named for David Hilbert (1862–1943), a German mathematician who worked with space-filling curves. The fractal line follows a mazelike pattern starting at the top left of the square and ending at the bottom left. As the levels increase, the design becomes more complex and fills more of the interior of the original space. After infinitely many iterations, the fractal fills the square.

Level 1 Level 2 Level 3 Level 4

As with many space-filling fractals, the Hilbert curve grows inward, while its outer boundary stays fixed.

Level 5

The fractal line segments of the Hilbert curve divide its plane into two spaces that are equal in area but very different in shape. When one of the areas is shaded, a pattern containing many symmetries emerges.

The quilt shows this kind of shading at Level 5 of the fractal design. To create the foundation of Level 5, the original square is divided into 1024, or 4^5, smaller squares. A solid fabric and complementary patterned fabric were used to make the quilt. The quilting for this design followed the fractal lines. Thus, the design on the front of the quilt emphasizes the "hook" shapes for which this fractal is known, but the back of the quilt shows the original mazelike fractal from which the design was created.

Designing Your Quilt

As with the Sierpiński curve, each level of the Hilbert curve is created starting with a new square. To draw the levels of the Hilbert curve, you will need several copies of a large square drawn on rectangular dot paper or graph paper. Drawing the levels involves bisecting segments, so use a side length that is a power of 2, such as 32 or 64.

Level 0 of this fractal is an empty square. Start with one copy of the square and follow these steps to form Level 1.

Step 1 Divide the square into 4 congruent smaller squares by connecting the midpoints of the opposite sides.

Step 2 Using a pencil, draw a square in the center of the original square. The length of the sides of the new square should be one-half the length of the original square, with the sides of the new square bisecting the sides of the 4 smaller squares. Erase the left side of the new square. This U-shape is the basis for all future levels of the Hilbert curve.

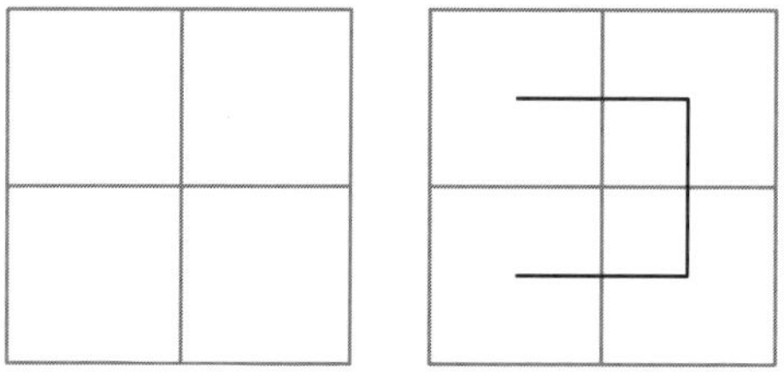

Level 1

Start with a second copy of the square and follow these steps to form Level 2.

Step 3 Repeat Step 1 to divide this square into 4 smaller squares.

Step 4 Divide each of the squares into 4 smaller squares using Step 1 as a guide. The figure should now have 16 small squares.

Step 5 Within each of the 4 squares formed in Step 3, draw squares using Step 2 as a guide. Working from left to right, row by row, erase one side of each square as follows: row 1 upper, left; and row 2 lower, left.

Step 6 Connect these three-sided figures with horizontal and vertical segments as shown in the rightmost drawing. The shape of Level 2 of the Hilbert curve is often called a *hook.*

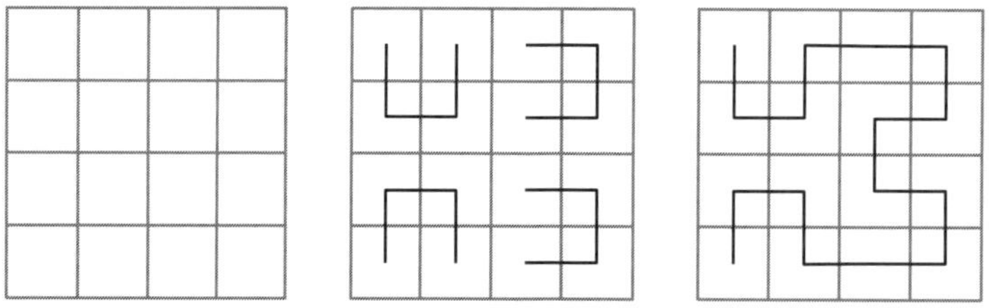

Level 2

Step 7 Recreate the foundation of 16 squares as for Level 2.

Step 8 Divide each of the 16 squares into 4 congruent squares. The figure should now have 64 squares.

Step 9 In each of the 16 larger squares, draw a hook as in Step 6.

Working from left to right, top to bottom, erase one side of each hook as follows:

Row 1 left, right, upper, left; row 2 upper, upper, lower, left; row 3 lower, lower, upper, left; and row 4 left, right, lower, left.

Step 10 Connect these three-sided hooks with horizontal and vertical line segments as shown in the rightmost drawing.

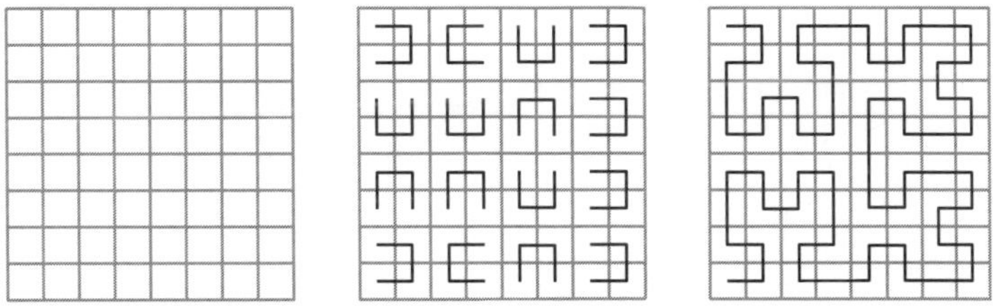

Level 3

Notice that Level 3 contains four copies of the hook seen in Level 2. The upper-right and lower-right hooks face the same direction as the hook in Level 2. The upper-left and lower-left hooks are rotated so the openings face away from each other. Also notice that the first U of each level alternates between opening upward and opening to the left.

Step 11 Color your design.

Making the Hilbert Curve Quilt

The finished Hilbert curve quilt measures 50 by 50 inches.

Materials

3 yards each of solid fabric and patterned fabric

$1\frac{1}{2}$ yards of batting

$1\frac{1}{2}$ yards of backing fabric

Binding fabric

Instructions

An illustration of Level 5 of the Hilbert curve can be found on page 74. Using the illustration as a model, draw the quilt pattern on graph paper, using one grid unit for the width of each "strip." The final drawing should measure 31 by 31 units. To make the quilt, you will create four 15-by-15-unit quadrants and then join them with 1-unit-wide connecting strips.

The following table gives the number, size, and fabric of the pieces needed for each quadrant of the quilt. Because all the pieces are 2 inches wide ($1\frac{1}{2}$ inches plus a seam allowance), you can use a rotary cutter to cut 2-inch strips. In addition to the listed pieces, you will need a 2-by-23-inch solid connector strip, a 2-by-23-inch print connector strip, and a 2-by-47-inch print connector strip.

Fabric	Size	Number of Pieces			
		Quad I Upper Right	**Quad II Upper Left**	**Quad III Lower Left**	**Quad IV Lower Right**
Solid	2″ × 2″	48	48	48	48
Print	2″ × 2″	22	22	27	27
Print	2″ × 3.5″	28	28	22	22
Solid	2″ × 5″	12	12	12	12
Print	2″ × 5″	8	8	7	7
Solid	2″ × 6.5″	—	2	—	—
Print	2″ × 6.5″	—	—	2	2
Print	2″ × 8″	2	2	1	1
Solid	2″ × 11″	3	3	3	3
Print	2″ × 12.5″	1	1	—	—
Solid	2″ × 23″	—	—	1	1

Create each quadrant separately, sewing each row together and then joining the rows. Use your drawing as a guide.

When the quadrants are completed, join quadrants II and III with a 2-by-23-inch print strip, and join quadrants I and IV with a 2-by-23-inch solid strip. Then connect the two halves with a 2-by-47-inch print strip.

Add a border if desired, and then add batting and a backing. To quilt this pattern, stitch along the path of the fractal design, outlining the dark design. Add a binding and a rod pocket for hanging.

ACTIVITY 1

Coastlines and Fractals

1. Use a compass, a ruler, and a copy of Worksheet 10, "Continental U.S. Map," to create the following table.
   a. Open your compass to 4 inches. Starting at Seattle, Washington, "walk" the compass around the border until you get back to Seattle, counting the number of "steps." When the last turn of the compass lands beyond Seattle, don't count it. Use your results as an estimate of the length of the map's border. For example, if it took five steps to go around the border, the estimate would be 4 inches $\times$ 5 steps $=$ 20 inches. Record your result in this table.
   b. Measure the border three more times, using a 2-inch compass opening, a 1-inch opening, and a $\frac{1}{2}$-inch opening. Record your results in this table.

Compass Opening	Number of Steps	Estimated Border Length (inches)
4 inches		
2 inches		
1 inch		
$\frac{1}{2}$ inch		

2. How do the border-length estimates change as the compass opening gets smaller?

WORKSHEET 10

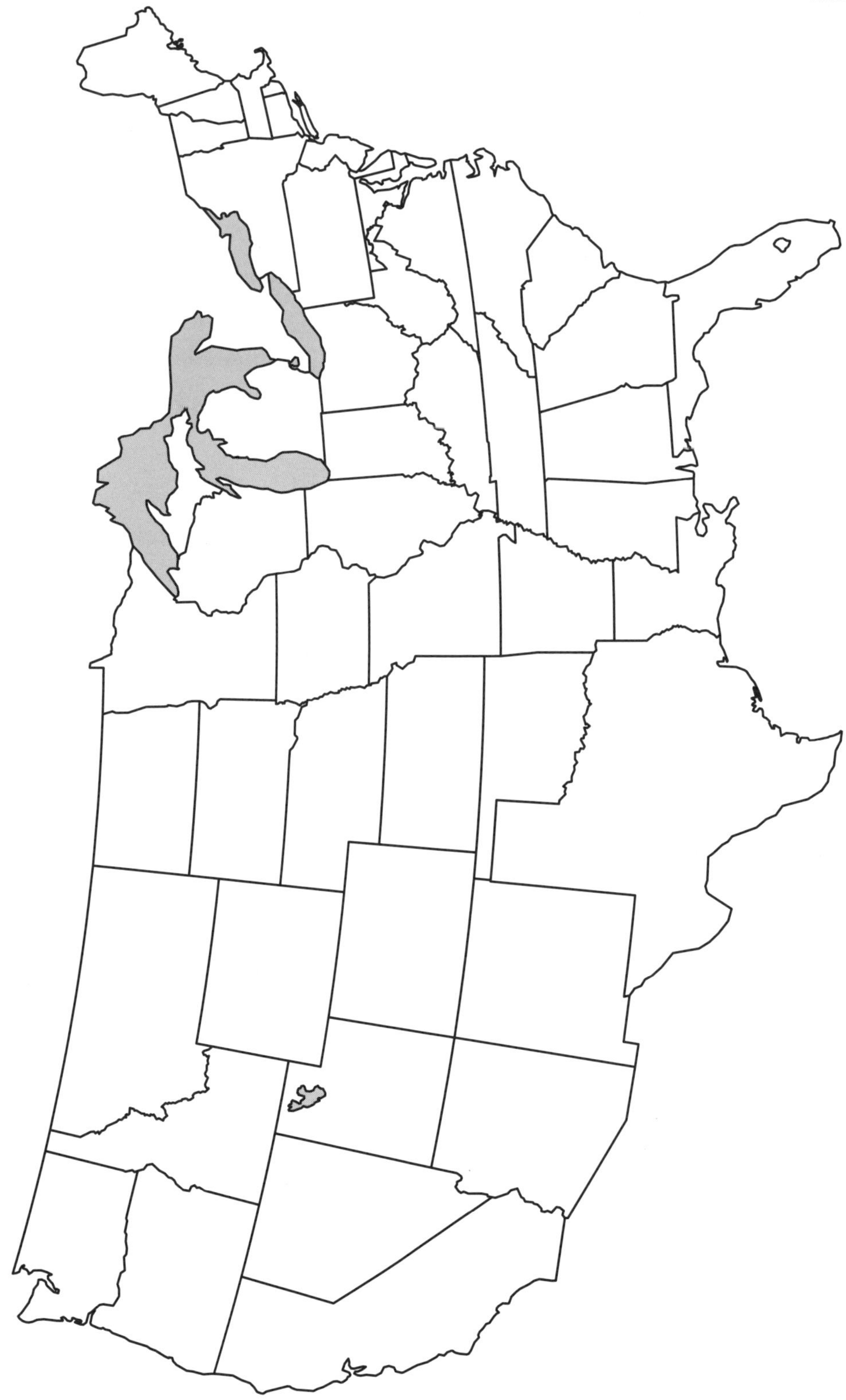

Exploring the Perimeter of the Koch Coastline

For this activity, you will need the fractal design you made in the "Designing Your Quilt" section (see page 49). If your design goes only to Level 3, extend at least part of it to Level 4 or use Worksheet 5.

1. Fill in the table for Levels 0 through 4 of the fractal. Consider the distance between two consecutive dots or grid lines to be 1 unit. Record your results in the table.

	Number of Segments	Length of Each Segment	Width	Height	Total Length (perimeter)
Level 0	1	81	81	0	81
Level 1	5	27	81	27	135 (27×5)
Level 2					
Level 3					
Level 4					

2. Why is the width the same for each level?
3. The height increases with each level. How large do you think it will eventually become? Why?
4. Calculate the given ratios

$$\frac{\text{perimeter of Level 1}}{\text{perimeter of Level 0}} \qquad \frac{\text{perimeter of Level 2}}{\text{perimeter of Level 1}}$$

$$\frac{\text{perimeter of Level 3}}{\text{perimeter of Level 2}} \qquad \frac{\text{perimeter of Level 4}}{\text{perimeter of Level 3}}$$

What pattern do you see in the results? Explain why this pattern occurs.

Creating a Reverse Snowflake

1. Follow these steps to create a reverse snowflake design.

   Step 1 Draw an equilateral triangle on a sheet of isometric dot paper. Use side lengths of 27 units or 81 units.

   Step 2 Divide each segment into three congruent parts.

   Step 3 Draw an equilateral triangle on the middle third of each segment so that the triangle is in the *interior* of the figure. Then erase the middle segment.

Level 1

   Step 4 Create each succeeding level by repeating Steps 2 and 3 on each new segment. Draw the fractal to Level 3.

2. Complete the tables for the Koch snowflake and for the reverse snowflake and compare the results.

Koch Snowflake

	Side Length	Number of Sides	Total Length (perimeter)
Level 0			
Level 1			
Level 2			
Level 3			

Reverse Snowflake

	Segment Length	Number of Segments	Total Length (perimeter)
Level 0			
Level 1			
Level 2			
Level 3			

3. Predict the total length of Levels 4 and 5 for each fractal.

Exploring Patterns in Linear Fractals

1. Look back at the Koch coastline and three-halves curve designs you drew, and complete each table.

Koch Coastline

	Number of Segments	Length of Each Segment	Total Length (perimeter)
Level 0	1	81	81
Level 1			
Level 2			
Level 3			

Three-Halves Curve

	Number of Segments	Length of Each Segment	Total Length (perimeter)
Level 0	1	64	64
Level 1			
Level 2			
Level 3			

2. Describe the patterns you see in the tables.
3. Predict the number of segments and total length for Levels 4 and 5.

Creating Your Own Fractal

The Koch coastline and three-halves curve each begin with a line segment, and the Koch snowflake begins with three segments as sides of an equilateral triangle. The next levels of these fractals are created by subdividing each segment and then replacing parts of the segment with new figures.

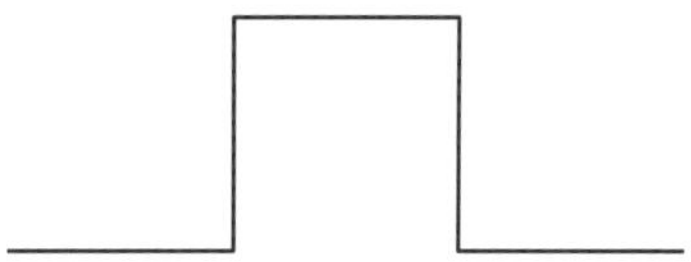

Koch coastline
Level 1

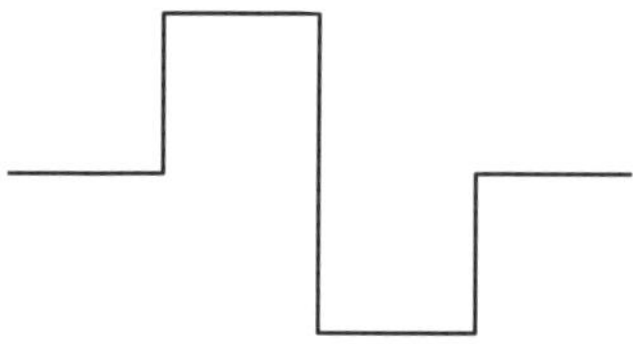

Three-Halves curve
Level 1

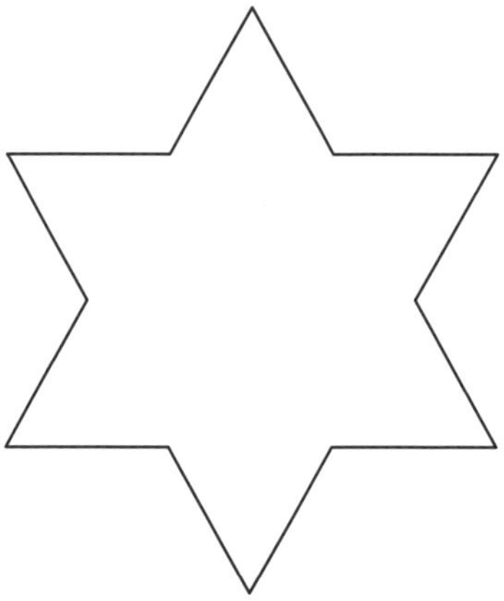

Koch snowflake
Level 1

Now create your own fractal that begins with a line segment. The levels of your fractal should be easy to draw using dot or grid paper as a guide. Draw Levels 0 through 3 of your fractal, and describe the steps you used to create it.

Fractal Dimension

The linear fractals in this section have dimensions between 1 and 2—somewhere between a line segment with one dimension (length) and a region with two dimensions (length and width).

To calculate the dimension of the three-halves curve, divide the logarithm of the *size factor* (the number of new segments that replace the initiator segment) by the logarithm of the *scale factor* (the number of segments the initiator segment is divided into).

A logarithm is an exponent. For the three-halves curve, the scale factor is 4 and the size factor is 8. You can use the LOG function on your calculator to calculate the dimension:

$$\frac{\log_x(\text{size factor})}{\log_x(\text{scale factor})} = \frac{\log_{10}8}{\log_{10}4} = 1.5$$

Or use base 2 instead of base 10: $\log_2 8 = 3$ ($2^3 = 8$) and $\log_2 4 = 2$ ($2^2 = 4$). So

$$\frac{\log_2 8}{\log_2 4} = \frac{3}{2}$$

Determine the fractal dimension of these other linear fractals.

1. Koch coastline

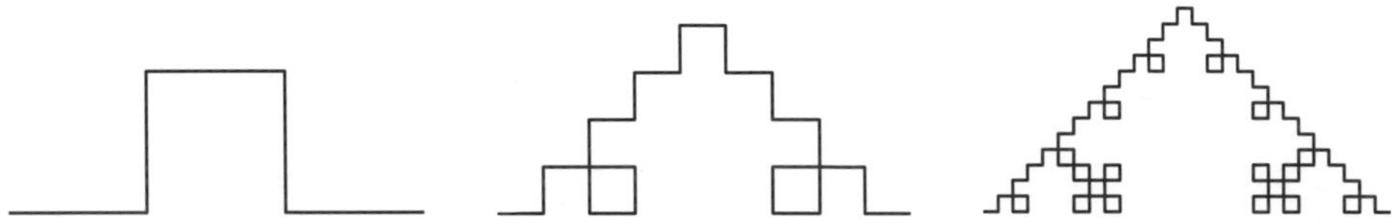

Level 1 Level 2 Level 3

2. Koch snowflake

Level 1 Level 2 Level 3

Symmetry in Fractals

A figure has *line symmetry* (also called *reflection symmetry*) if you can fold it in half along a line, called the *line of symmetry,* so that the two halves match exactly. A figure can have more than one line of symmetry.

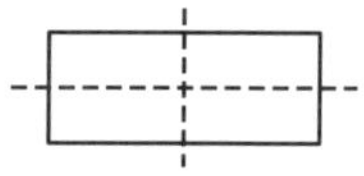

Two lines of symmetry

A figure has *rotation symmetry* if you can rotate it less than a full turn about a centerpoint so that it looks the same as it did in its original position.

Rotational symmetry

1. Find all the lines of symmetry for these fractal designs. How many did you find for each?

   a. Sierpiński carpet

   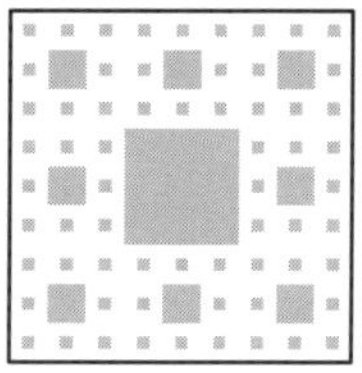

   b. Pythagorean tree

   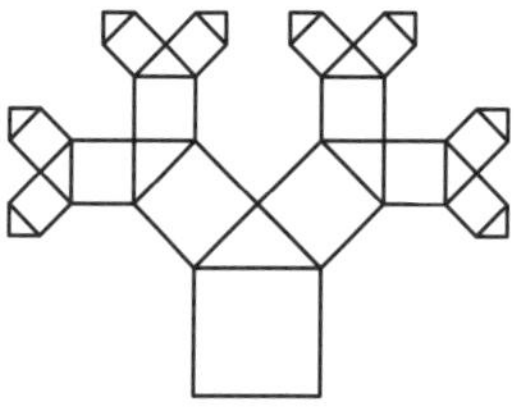

   c. Sierpiński triangle

   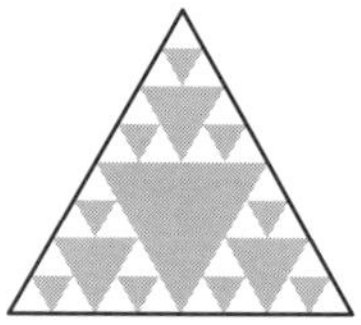

   d. Koch coastline

   e. Koch snowflake

   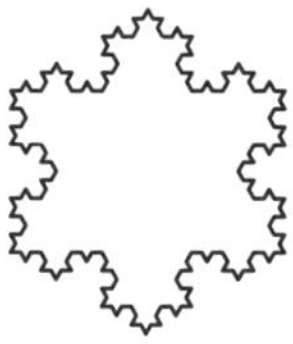

   f. Sierpiński curve

   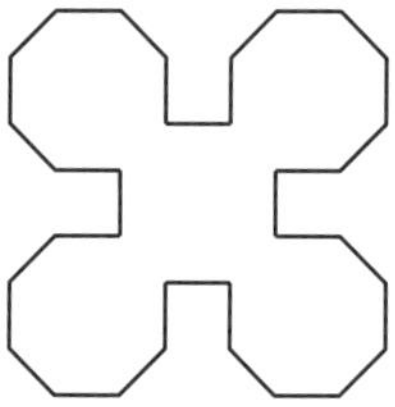

2. Which of the fractal designs in part 1 has point symmetry?
3. Does the three-halves curve have line symmetry? Does it have point symmetry?
4. How can you determine whether a fractal will have line symmetry or point symmetry based on the steps involved in creating it?

Exploring Patterns in the Hilbert Curve

The foundation for each level of the Hilbert curve includes more squares than the level before. The number of U-shapes and the number of hooks also increase with each level.

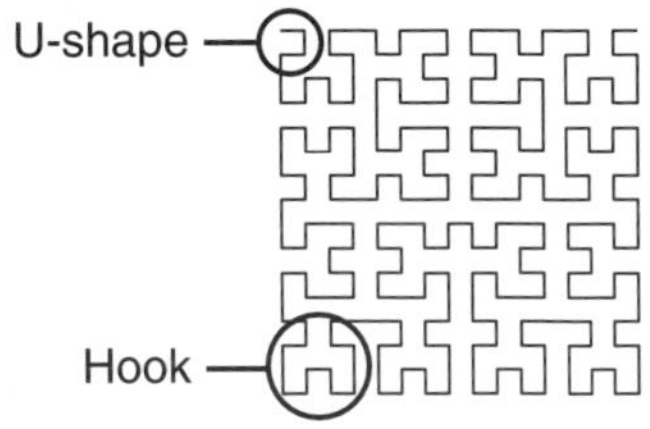

Level 4

1. Complete the table.

	Number of Smallest Squares	Number of U-Shapes	Number of Hooks
Level 0	0	0	0
Level 1	4	1	0
Level 2	16		
Level 3			
Level 4			
Level 5			

2. Describe the pattern in each column.
3. Predict the number of squares, U-shapes, and hooks in Level 6.
4. The nonzero entries in all the columns can be written as consecutive powers of the same base. What is the base? Why does it make sense that this number is the base?

Mazes and Friezes

Some linear fractals, such as the Hilbert curve, have a mazelike appearance. People have been creating maze designs for hundreds of years. Maze patterns have been found in mosaic floor designs dating back to the first century B.C. The study of mazes is part of the branch of mathematics called *topology.* Topologists categorize mazes according to the position of the paths and the overall shape of the maze.

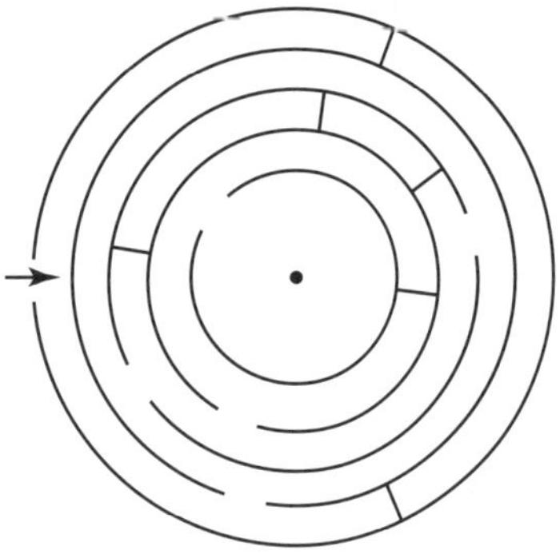

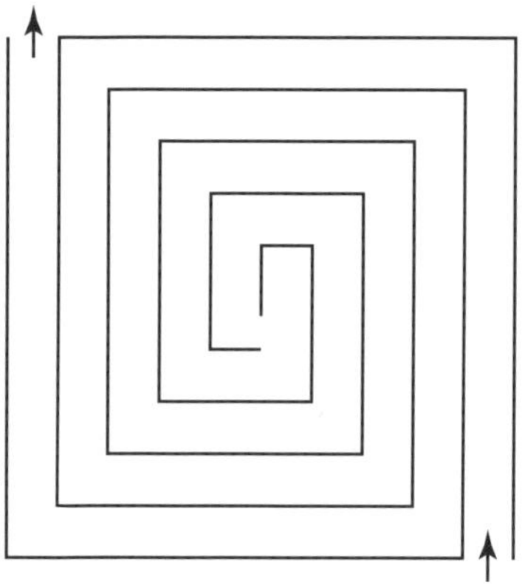

A *frieze* (pronounced "freeze") is a design in which a pattern is repeated many times in a horizontal or vertical direction. Friezes are commonly found on wallpaper, fabric, and borders of artwork. A frieze is formed by sliding and reflecting or rotating duplicates of an original image along a straight-line path.

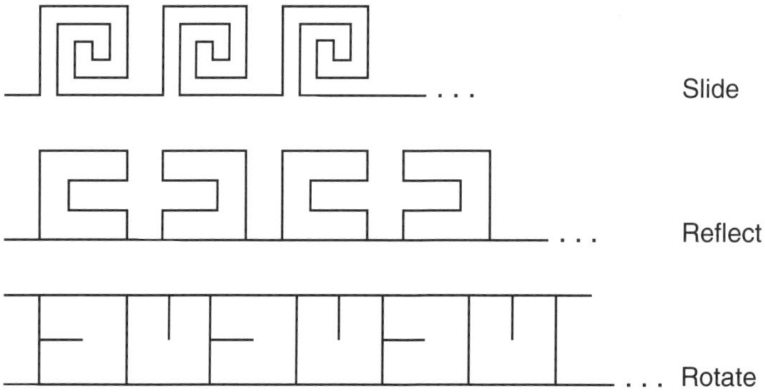

1. Many friezes consist of repetitions of a maze. To create your own frieze with a maze, you will need a sheet of blank paper, a sheet of tracing paper, and a pencil. In a typical maze, the path is the space between the line segments. However, in creating a frieze, think of the maze as the line segments themselves.
   a. On the blank sheet of paper, lightly draw a straight, horizontal line segment to use as a guide in creating the frieze.
   b. With a pencil, carefully trace the maze shown here onto the tracing paper.

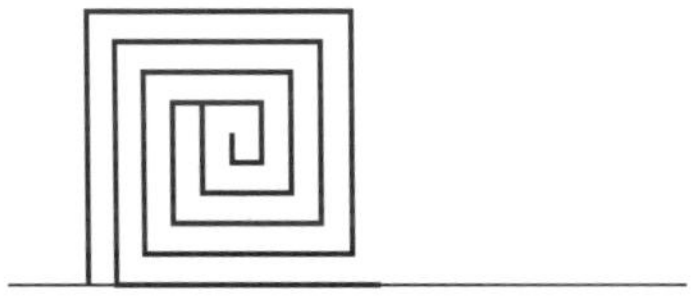

   c. Turn the tracing paper over, and trace over the maze lines with a pencil.
   d. Turn the tracing paper over again so the original tracing is face up. Place the tracing on the left end of the segment you drew in part a so the bottom of the maze is aligned with the segment.

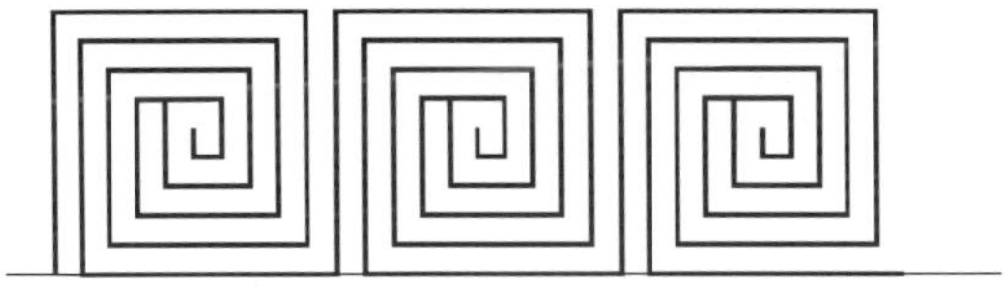

   e. Trace over the maze so the pencil lines on the other side of the tracing transfer to the paper.
   f. Now slide the tracing to the right so the exit point from the first maze lines up with the entrance point of the second. Trace the maze in its new position. (You may need to retrace the maze on the back of the tracing paper.)

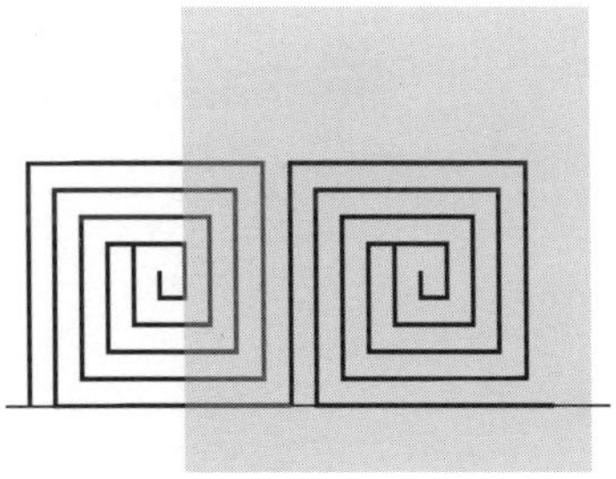

   g. Continue this process until your frieze is completed. Darken the segments of the final design with pen or pencil if necessary.
2. Now create your own frieze. Create a maze or other design, and then follow the steps in part 1 to create a frieze.

ACTIVITY 10

Research Topics

Research any of the topics listed below. You might search the Internet to find information about fractals and other topics in this chapter.

chaos theory
fractal
fractal dimension
frieze
Hausdorff dimension
Hilbert, David
Hilbert's curve
Koch coastline
Koch snowflake
linear fractal
Mandelbrot, Benoit
maze
Minkowski's curve
Minkowski's sausage
Sierpiński, Waclaw
topology
von Koch, Helge

Teacher Notes

Most students find studying fractals and creating fractal designs interesting and exciting. As they draw the designs on rectangular dot paper (Koch coastline and the three-halves curve) or isometric dot paper (Koch snowflake), students can observe how the fractal grows and becomes more complex. When students create their own fractal designs, encourage them to keep the changes simple, thinking ahead to higher levels where transformations must be done many times. After students have created or re-created a fractal design, they can add color.

Before students work on their own quilt designs, you might discuss with them the characteristics of fractals (see pages 43–50), illustrating the important ideas with an example. You might also discuss the idea of noninteger dimension if you plan to do Activity 6. The concept of fractal dimension requires your students to be open to new interpretations of traditional concepts.

For the Koch coastline and Koch snowflake, the levels are created by replacing the middle third of each segment with a new shape. The three-halves curve is slightly more complex because it requires two replacements and it twists and turns as it grows. Students may find it helpful to turn the paper by 90 degrees as they perform each transformation.

The Hilbert curve and the Sierpiński curve are interesting visually, but are difficult to describe and create. Students may be satisfied to just color the fractal designs provided on the worksheets. If they want to create their own versions of these fractals, encourage them to explore such fractals as the three-halves curve, Koch coastline, and Koch snowflake.

The Sierpiński and Hilbert curves are similar in their construction. For both curves, each level starts with a new square rather than building on the previous level. Students will need several copies of the square for each fractal design to create higher levels of the design (see page 179–185). Students may want to practice constructing the first few levels.

In Activity 1, students use a compass to measure the border of the continental United States. They will find that smaller compass openings give greater border lengths. This activity relates to Mandelbrot's idea that the length of a coastline or border is, in a sense, infinite—using a smaller unit of measure will always give a greater length. Results might be more dramatic with a larger map.

In Activities 2–4, students investigate patterns in the Koch coastline, Koch snowflake, and three-halves curve. In Activity 3, students create a reverse snowflake by performing the Koch snowflake transformations on the other side of each segment. In Activity 5, students create their own fractal, starting with a line segment. Activity 6 gives students practice in calculating the dimensions of linear fractals. Activity 7 reinforces the concept of symmetry. In Activity 8, students look at patterns in the Hilbert curve and discover how the patterns can be described in terms of exponential powers.

Activity 9 provides practice in transformational geometry, as students slide a maze along a line segment to create a frieze design. This activity illustrates the relationship between art and mathematics, and the use of mathematics in the creation of decorative objects throughout history.

Answers and Comments

Activity 1

A compass and a ruler are needed for this activity. Before starting the map activity, discuss with students the work of the British scientist Lewis F. Richardson (1881–1953). Richardson found that reports of the lengths of coastlines and of borders between countries varied greatly, and surmised that the reported length depended on the scale of measurement used. Benoit Mandelbrot (b. 1924) became interested in Richardson's work and began to investigate the geometry of borders and coastlines. Mandelbrot found that, in a sense, a coastline or border is infinitely long—no matter what measurement is given, a greater measurement can always be found by using a smaller unit of measure.

1. Answers may vary, especially for smaller compass openings.

Compass Opening	Number of Steps	Estimated Border Length (inches)
4 inches	5	20
2 inches	11	22
1 inch	24	24
$\frac{1}{2}$ inch	58	29

2. As the compass opening gets smaller, the estimated border length increases.

Activity 2

1.

	Number of Segments	Length of Each Segment	Width	Height	Total Length (perimeter)
Level 2	25	9	81	36 (27 + 9)	225 (9 × 25)
Level 3	125	3	81	39 (36 + 3)	375 (3 × 125)
Level 4	625	1	81	40 (39 + 1)	625 (1 × 625)

2. Because all the transformations occur above the initiator segment, the width stays at 81 units for each level.
3. After reaching a height of 40, the height will increase in small fractional increments ($\frac{1}{3}, \frac{1}{9}, \frac{1}{27}$, and so on). The height appears to approach $\frac{81}{2}$.
4. $\frac{\text{perimeter of Level 1}}{\text{perimeter of Level 0}} = \frac{135}{81} = \frac{5}{3} = 1.\overline{6}$

   $\frac{\text{perimeter of Level 2}}{\text{perimeter of Level 1}} = \frac{225}{135} = \frac{5}{3} = 1.\overline{6}$

   $\frac{\text{perimeter of Level 3}}{\text{perimeter of Level 2}} = \frac{375}{225} = \frac{5}{3} = 1.\overline{6}$

   $\frac{\text{perimeter of Level 4}}{\text{perimeter of Level 3}} = \frac{625}{375} = \frac{5}{3} = 1.\overline{6}$

   The ratios are all equal to $\frac{5}{3}$. This is because each level has 5 times as many segments as the previous level, and each segment is $\frac{1}{3}$ as long as the segments in the previous level. The total length of each level is $\frac{5}{3}$ times the length of the previous level.

Activity 3

1.

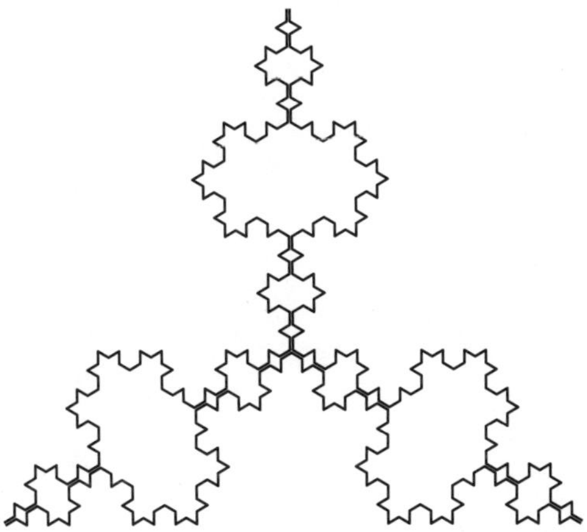

2. The entries are the same in both tables. The total length of Level 4 is 256; Level 5 is $341\frac{1}{3}$; each level is $\frac{4}{3}$ times the previous level.

Koch Snowflake

	Side Length	Number of Sides	Total Length (perimeter)
Level 0	27	3	81
Level 1	9	12	108
Level 2	3	48	144
Level 3	1	192	192

Reverse Snowflake

	Segment Length	Number of Segments	Total Length (perimeter)
Level 0	27	3	81
Level 1	9	12	108
Level 2	3	48	144
Level 3	1	192	192

Activity 4

1. **Koch Coastline**

	Number of Segments	Length of Each Segment	Total Length (perimeter)
Level 0	1	81	81
Level 1	5	27	135
Level 2	25	9	225
Level 3	125	3	375

Three-Halves Curve

	Number of Segments	Length of Each Segment	Total Length (perimeter)
Level 0	1	64	64
Level 1	8	16	128
Level 2	64	4	256
Level 3	512	1	512

2. Possible answers:

   Koch coastline: The number of segments in each level is 5 times the number in the previous level. The length of each segment is $\frac{1}{3}$ the length of each segment in the previous level. The total length of each level is $\frac{5}{3}$ the total length of the previous level.

   Three-halves curve: The number of segments in each level is 8 times the number in the previous level. The length of each segment is $\frac{1}{4}$ the length of each segment in the previous level. The total length of each level is twice the total length of the previous level.

3. **Koch Coastline**

	Number of Segments	Length of Each Segment
Level 4	625	625
Level 5	3125	$1041\frac{2}{3}$

Three-Halves Curve

	Number of Segments	Length of Each Segment
Level 4	4096	1024
Level 5	32,768	2048

Activity 5

Answers will vary. Check for iterations and self-similarity.

TEACHER NOTES

Activity 6

1. Koch coastline: $\frac{\log 5}{\log 3} = 1.4649\ldots$
2. Koch curve: $\frac{\log 4}{\log 3} = 1.2618\ldots$

Activity 7

1. a. 4 b. 1 c. 3
   d. 1 e. 6 f. 4
2. The Sierpiński carpet, Sierpiński triangle, Koch snowflake, and Sierpiński curve have point symmetry.
3. The three-halves curve does not have line symmetry. It does have point symmetry (it can be rotated 180 degrees about its centerpoint).
4. Possible answer: If the Level 1 shape is symmetric and the rule for creating a new level leaves the shape symmetric, then each level will be symmetric.

Activity 8

1.

	Number of Smallest Squares	Number of U-Shapes	Number of Hooks
Level 0	1	0	0
Level 1	4	1	0
Level 2	16	4	1
Level 3	64	16	4
Level 4	256	64	16
Level 5	1024	256	64

2. Number of squares: There is one square for Level 0. The number of squares in each succeeding level is four times the number in the previous level.

   Number of U-shapes: There are zero U-shapes in Level 0 and one U-shape in Level 1. The number of U-shapes in each succeeding level is four times the number in the previous level.

   Number of hooks: There are zero hooks in Levels 0 and 1 and one hook in Level 2. The number of hooks in each succeeding level is four times the number in the previous level.

3. Squares: $4 \times 1024 = 4096$; U-shapes: $4 \times 256 = 1024$; hooks: $4 \times 64 = 256$

4. All the entries can be written as powers of 4. This makes sense because at each level, each square in the previous foundation is divided into four squares. So the number of squares is multiplied by 4 with each level. Because the U-shapes and hooks are built on the squares, the number of each of these shapes is also multiplied by 4 with each level.

Activity 9

Frieze designs will vary.

The Spatial Quilts

THE QUILT DESIGNS IN THIS SECTION—Tumbling Block Pyramid, World of Geometry, and Clifford Torus—are examples of interpreting three- and four-dimensional mathematical objects in two- or three-dimensional space.

Throughout history, artists, architects, scientists, mathematicians, and others have studied the three dimensions we live in, as well as those beyond the third. They have confronted the problem of representing three-dimensional objects within the confines of a two-dimensional object, such as a canvas, blueprint, chalkboard, graph paper, or computer screen. They have also created representations of three-dimensional projections (shadows) of four-dimensional objects.

The Tumbling Block Pyramid quilt represents, on a two-dimensional surface, a three-dimensional pyramid of cubes. Only three of the six faces of the cubes on one side of the pyramid are visible. You must mentally visualize the hidden faces and cubes

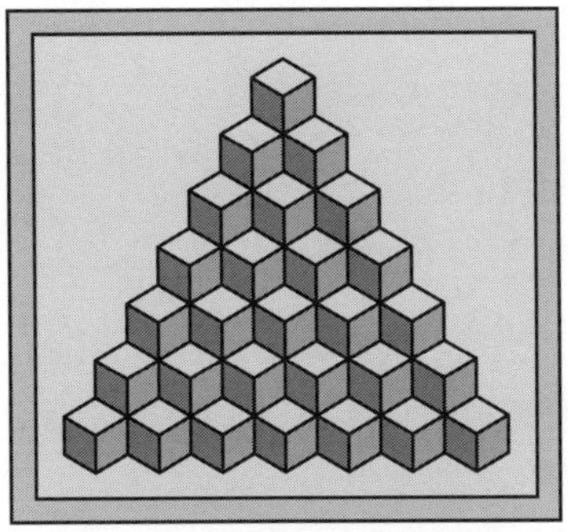

The Tumbling Block Pyramid Quilt

The World of Geometry Quilt

The Clifford Torus Quilt

in order to see the pyramid. The three color values used in the fabric help define the dimensions of the cubes by suggesting shadows.

The quilt shows an *isometric* view of the cubes. An isometric image gives the illusion of depth while preserving the relative dimensions of the object pictured. For example, in an isometric drawing of a cube, all the edges are the same length, just as in an actual cube.

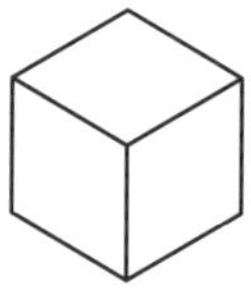

Rendered as a two-dimensional design, the quilt pattern consists of a tessellation of rhombuses with two different centers of threefold rotational symmetry. At one center of rotation, six 60-degree angles meet; at the other center, three 120-degree angles meet.

Three-dimensional objects can also be represented using *perspective*. In a perspective drawing, parallel lines converge to a *vanishing point*, creating the illusion of depth and distance. Unlike an isometric drawing, a perspective drawing does not preserve the relative dimensions of some objects. In this perspective drawing of a cube, not all edges are drawn to the same length.

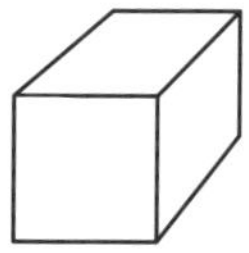

The World of Geometry quilt is a three-dimensional globe. Mathematicians Stuart Levy and Brad Barber created this design while investigating ways to represent Earth on a polyhedron's surface. They designed a globe made of 1620 triangles; but during the design process, they also created patterns for globes made from 20, 50, and 175 triangles.

Using triangular shapes to create polyhedra is not a new process. The ancient Egyptians used the rigidity of triangles in constructing the pyramids. Three of the five Platonic solids have triangular faces. The engineer Buckminster Fuller (1895–1983) used rigid triangles to design buildings in the shape of geodesic domes (hemispheres in which either five triangles or six triangles meet at a vertex). Architects and builders continue to use triangles in construction.

In the seventeenth century, projective and coordinate geometry became important tools for linking representations of physical objects to a mathematical structure. The French mathematician Gerard Desargues (1591–1661) developed ideas in projective geometry by asking what different perspective drawings of an object had in common. For example, there are several ways to see the shadow of a cube's vertices and edges. Three ways to represent a cube on a two-dimensional surface are: centered on a vertex, centered on an edge, and from above.

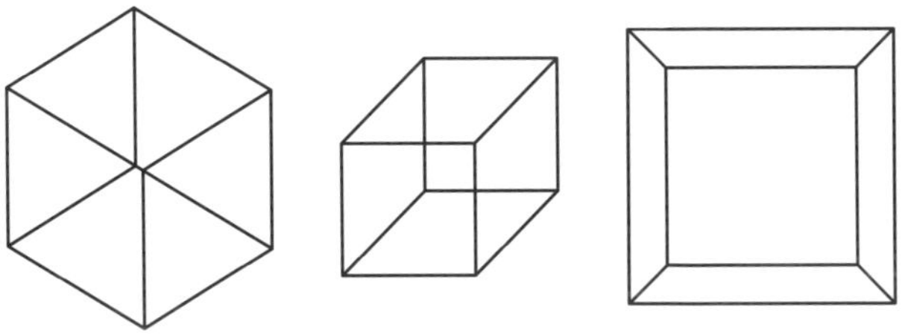

In each projection, there are 12 edges and 8 vertices.

René Descartes (1596–1650) helped to develop the *Cartesian coordinate system,* two perpendicular number lines (axes) that divide the plane into four quadrants, or three perpendicular axes that divide space into eight sectors. Descartes and Pierre de Fermat (1601–1665) developed the branch of mathematics called *analytic geometry,* which uses algebra and the Cartesian coordinate system to study geometric figures.

The Clifford torus is one view of a four-dimensional *hypersphere* projected in three dimensions on a two-dimensional computer screen. The Clifford torus is "banded" to allow us to look at the "front" and "back" of the projection. It shows just one of many ways of looking at the shadow. (Dr. Banchoff's Web site shows other views. It can be reached from **www.keymath.com/quilts**.)

The Clifford torus is one twentieth-century outcome of the nineteenth-century's interest in higher dimensions and non-Euclidean spaces. (Two-dimensional Euclidean space is a plane. The surface of a sphere is a two-dimensional non-Euclidean space that can be studied on the surface of a sphere such as the Lénárt sphere.)

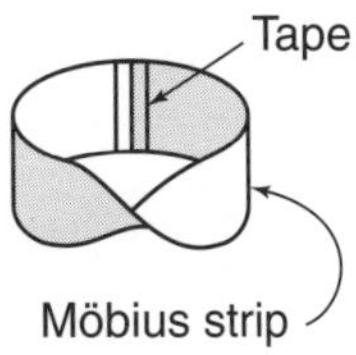

The Möbius strip, invented by the German mathematician August Möbius (1790–1868) in 1840, is a two-dimensional surface with only one side and one edge. Any point on the surface can be reached without going over the edge. In two dimensions it appears to have just one side, but from three dimensions we can see it has a twist. Extending this image helps us imagine a three-dimensional space without boundary. The book *Exploring the Shape of Space* by

Jeffrey Weeks and the video *Shape of Space* explore extensions of the Möbius strip to three-dimensional spaces without boundaries, which in four dimensions would be seen to have a twist.

Geometers, scientists, philosophers, and artists have often explored dimensions beyond the third through metaphor, analogy, and more formal methods of coordinate geometry. Thomas Banchoff (b. 1938) and other present-day mathematicians use computer graphics to represent surfaces, such as the Clifford torus, that have more than three dimensions. These computer-generated images can help us grasp new concepts.

The many symmetries of the Clifford torus in four-dimensional space are studied in geometry and topology. As with many discoveries, it was originally thought that the torus had few, if any, applications. However, recently it has been used in physics to study the position and velocity of pendulums.

Computer images and ideas of four dimensions indicate that dimension is not limited to length, width, and height. Einstein's work with relativity has led us to an exploration of time as a fourth dimension. Theoretical physicists investigating the basic structure of the universe study eleven-dimensional strings.

The Tumbling Block Pyramid Quilt

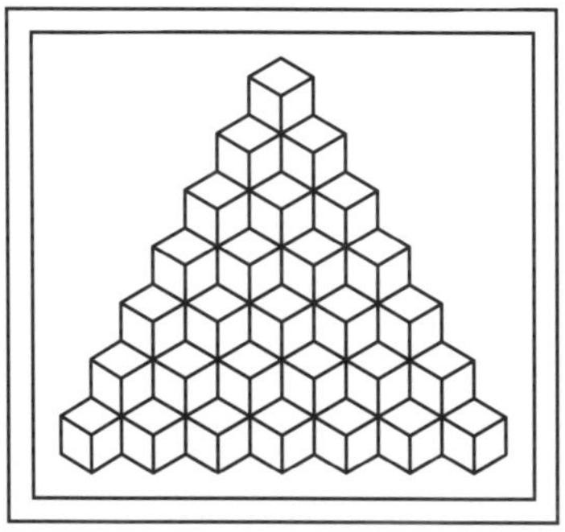

Tumbling blocks is a traditional quilt pattern dating back to the eighteenth century. The design is constructed by tessellating (covering a surface with no gaps or overlaps) one shape—a rhombus. The placement and colors of the rhombuses create the illusion of a three-dimensional pyramid built from cubes.

The measures of the angles surrounding any vertex in a tessellation must have a sum of 360 degrees. The rhombuses used in the quilt have two 60-degree angles and two 120-degree angles. At the vertices the rhombuses intersect in two ways. At some vertices, six 60-degree angles meet; at others, three 120-degree angles meet. The tessellation of rhombuses has two different centers of threefold revolution.

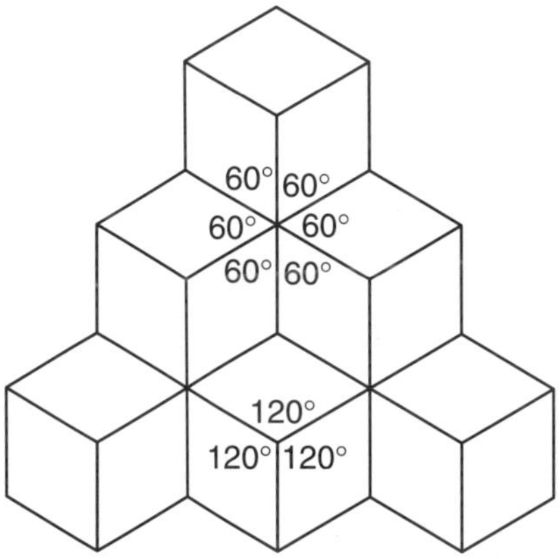

Three groups of colors—dark ones on the top, light ones on the right, and medium ones on the left—are used to enhance the illusion of cubes in the quilt.

Designing Your Quilt

To create the tumbling block pyramid, make a template for the rhombus shape, and then trace it many times.

Step 1 Cut from a sheet of cardboard or plastic a rhombus similar to the one pictured. Check to be sure that the sides of your rhombus are all the same length (about $1\frac{1}{4}$ inches) and that the angles measure 60 degrees and 120 degrees.

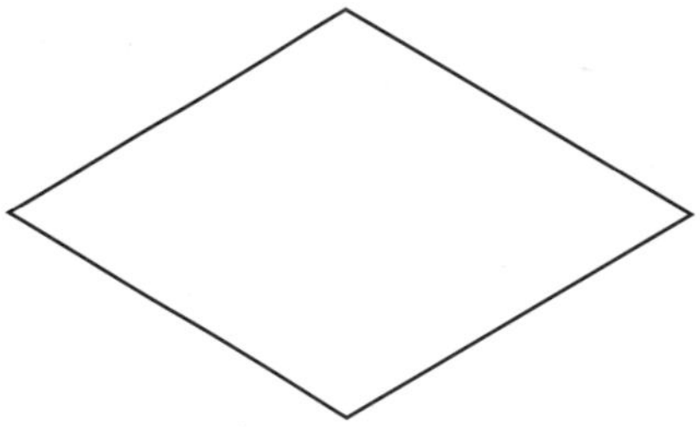

Step 2 Near the top of a sheet of blank paper, trace the template three times to create this picture of a cube. This will be the top row of the pyramid.

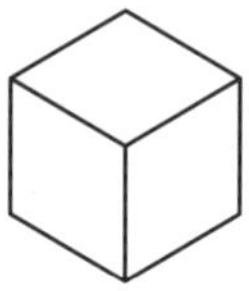

Step 3 The second row contains two cubes. Trace the template six more times to create this row.

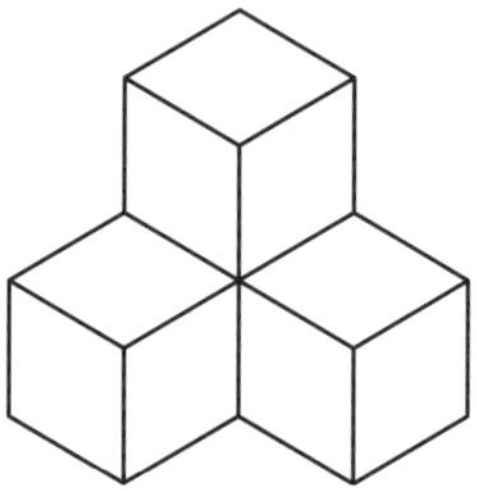

Step 4 Continue tracing the template to create the remaining rows of the pyramid. Each row should have one more cube than the row above it.

Step 5 Color your design. To create the illusion of a three-dimensional pyramid, use dark colors in the same positions in each cube and do the same for medium colors and light colors.

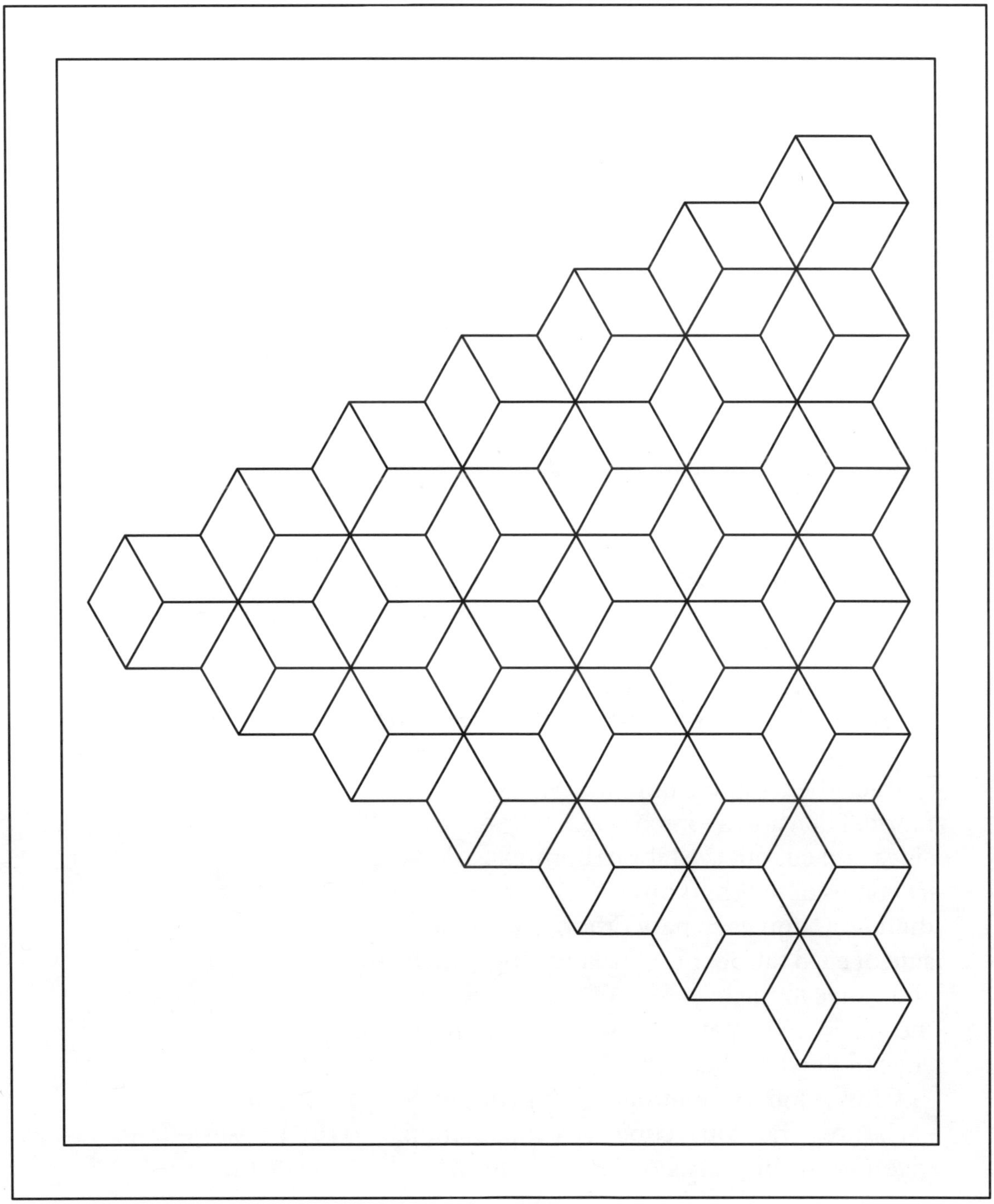

Making the Tumbling Block Quilt

The Tumbling Block quilt measures 30 inches wide by 24 inches high.

Materials

Small pieces of fabric for the rhombuses (28 dark, 28 medium, and 28 light)

$\frac{3}{4}$ yard of background fabric

$\frac{1}{4}$ yard of border fabric

$\frac{1}{2}$ yard of binding fabric

$\frac{3}{4}$ yard of batting

1 yard of backing fabric

Cardboard or plastic for a rhombus template

Instructions

This quilt is made using English paper piecing—a paper copy of each rhombus is used to secure the fabric rhombuses as they are whipstitched together.

Create a cardboard or plastic template of a rhombus with side length of $1\frac{1}{4}$ inches and angles of 60 degrees and 120 degrees. (See page 101.)

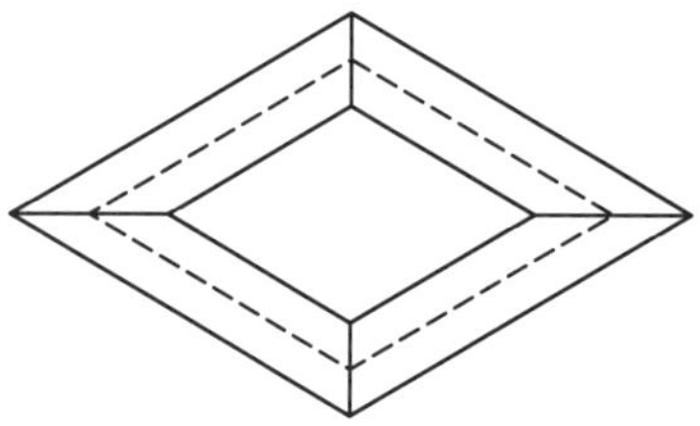

Trace the template onto the wrong side of the fabric. Then add a $\frac{1}{4}$-inch seam allowance on all sides and cut out the fabric rhombuses. Make 84 copies of the rhombus on freezer paper and cut them out. Iron each paper rhombus to the wrong side of each fabric rhombus. Fold the $\frac{1}{4}$-inch seam allowance to the back side of the paper, and baste the fabric to the paper about $\frac{1}{8}$ inch from the outside edge of the rhombus.

Arrange the 84 rhombuses in a pyramid design. Assemble each three-rhombus "cube" by whipstitching two rhombuses together—using very small stitches from the back side along the outer edge of each piece—and then joining the third rhombus to the first two.

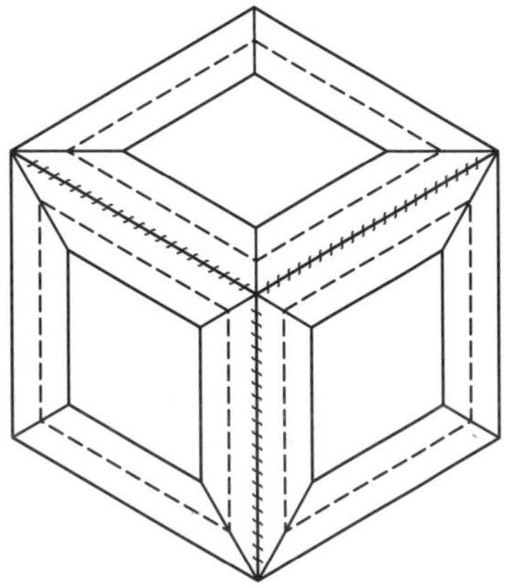

When you have finished all 28 cubes, whipstitch them together, one row at a time, to form a pyramid.

Appliqué the pyramid to the background fabric. Then turn the quilt over and carefully cut away the background fabric from the pyramid section, leaving a $\frac{1}{4}$-inch seam allowance. Remove the basting stitches and paper.

Add a border if desired. Baste the quilt top to the batting and backing. Quilt the layers together by stitching straight lines from the sides of the pyramid to the edge of the quilt. Add a binding and a rod pocket.

The World of Geometry Quilt

The World of Geometry quilt is an icosahedral globe made from 20 fabric triangles. The pattern, created by Stuart Levy and Brad Barber, was part of a project with the Geometry Center in Minneapolis, Minnesota, to build a 1620-triangle globe. A globe with a triangular grid, in which the regions are the same size, can be used to study topics such as biological diversity. In contrast, regions bounded by the standard latitude and longitude lines are smaller near the poles than at the equator.

Designing Your Quilt

Globe patterns for the 20-, 50-, 80-, and 121-triangle globes are available at Stuart Levy's home page, which can be reached from www.keymath.com/quilts.

Or you could make your own pattern from an inflatable plastic globe. After inflating the globe, locate the equator, the North Pole, and the South Pole. Use a permanent marker to mark both poles and highlight the equator. Surround both the North Pole and the South Pole with five equilateral triangles. The five triangles must have a common vertex at the poles. The triangles will extend $\frac{1}{3}$ of the way down from the North Pole to the South Pole.

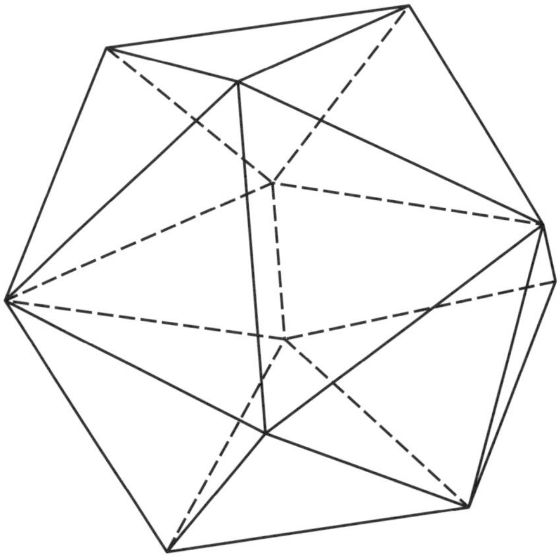

Mark a band of ten triangular regions around the middle of the earth. Use the tetrahedral globe on page 106 as a guide.

Once the 20 triangles are determined, cut out each one carefully, paying close attention to the order in which you will reassemble them. Lay the pieces out in order and transfer the patterns onto heavy paper or fabric allowing a $\frac{1}{4}$-inch border for gluing flaps (or for a seam allowance). Color the individual pieces, then glue them together to form an icosahedron.

Making the World of Geometry Quilt

Materials

3 yards of cotton fabric

Heavy fusible interfacing

3 yards of batting

10 skeins of embroidery thread

One large, flat button

One heavy balloon or batting

Instructions

Obtain patterns for the triangular pieces at **www.keymath.com/quilts.** Wash the material in hot water, dry it, and iron it completely flat. As you place the patterns for the 20 (or 50) triangles on the cloth, envision the way the triangles will fit together on the globe. Keep the grain of the fabric on adjacent pieces going in the same direction.

Place each triangle on the heavy fusible interfacing. Iron the interfacing to the wrong side of each triangle, and trim the excess interfacing so that only a little will remain in the seam allowance. Cut out batting in the shape of each piece and pin it to the wrong side of the triangle. You now have a sandwich, with the pattern on top, material and interfacing in the middle, and batting on the bottom. Use a sewing machine to stitch around each triangle just outside the seam line. This will keep all the layers fixed in place as you work.

Sew the latitude and longitude lines on each triangle. Embroider the outline of the continents, sewing right through the pattern. In the quilt shown, the continents, states, and key bodies of water were outlined. You can add as much detail as you like. Once all the triangles are embroidered, tear away the paper pattern.

To assemble the quilt, start by piecing together the South Pole. Then work your way up to the North Pole. Sew a large, flat button on the inside of the North Pole to stabilize the sphere while it is hanging. Leave a small hole at the North Pole to insert batting if the globe is small or a balloon for large globes. Insert the balloon and inflate it. Sew the last seam by hand, being very careful not to pop the balloon. If the balloon deflates, replace it with a new balloon.

The Clifford Torus Quilt

The Clifford Torus quilt is named for the English mathematician William Clifford (1845–1879). The quilt design was inspired by the computer graphics images of Dr. Thomas F. Banchoff. Banchoff's image shows a continuous rainbow-colored band around the torus. It can be rotated to show different projections or shadows of the four-dimensional object. Here is another view of the Clifford torus.

Designing Your Quilt

In his book *Beyond the Third Dimension: Geometry, Computer Graphics, and Higher Dimensions,* Banchoff shows seven views of the Clifford torus. Dr. Banchoff's home page (see **www.keymath.com/quilts**) shows different projections of the torus that can be rotated to see the projections from different angles. Start with a sketch from Banchoff's book or Web site, or use Worksheet 12 and add colors.

You could create other quilt designs to represent other dimensions, such as a hypercube. Or you might create a design inspired by the satire *Flatland,* a book about life in a two-dimensional world.

Designing Your Quilt

WORKSHEET 12

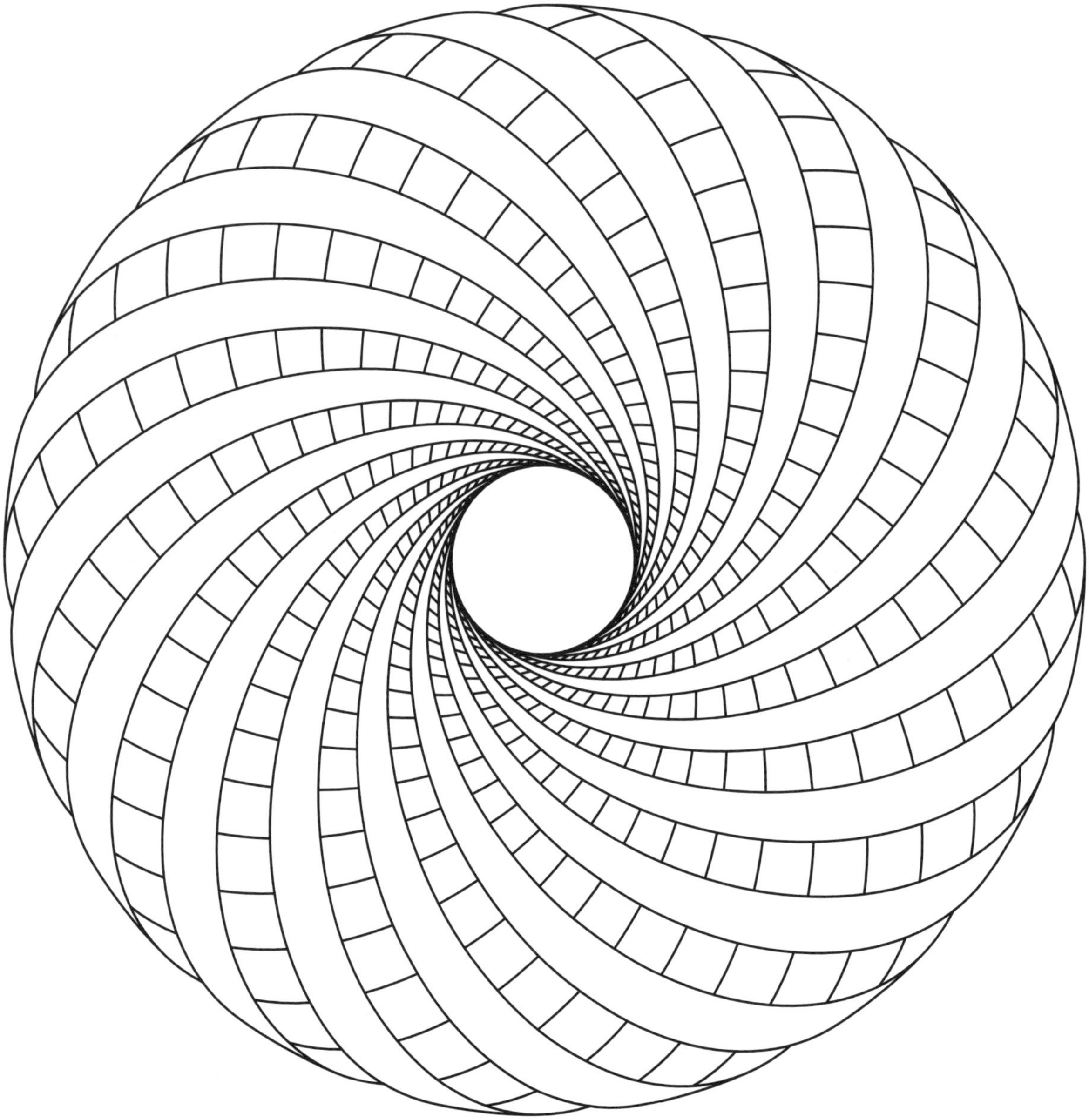

Making the Clifford Torus Quilt

The Clifford Torus quilt measures 40 inches by 40 inches.

Materials

$2\frac{1}{2}$ yards of fabric for the front of the quilt

$1\frac{1}{2}$ yards of backing material

$\frac{1}{8}$ to $\frac{1}{4}$ yard of fabric for each of the 16 colored bands

Freezer paper

Instructions

Draw the Clifford torus on the unsticky side of the freezer paper by enlarging the design from Worksheet 12. Trace the same image onto the background fabric.

Cut each small piece out of the freezer paper. Press the sticky side of the paper to the fabric and cut out the fabric, adding a $\frac{1}{4}$-inch seam allowance. Press the seam allowance to the back side of the paper, then appliqué the piece to the background. Just before you finish sewing a piece to the background, remove the paper.

Layer the quilt with the batting and backing. Start the quilting by outlining the very center of the quilt. For the outside portions of the quilt, continue the circular stitching that outlines each band of the torus to the edge. Add a binding and a rod pocket.

ACTIVITY 1

Isometric Drawing

1. Using isometric dot paper, draw and color or shade each of these designs.

A.

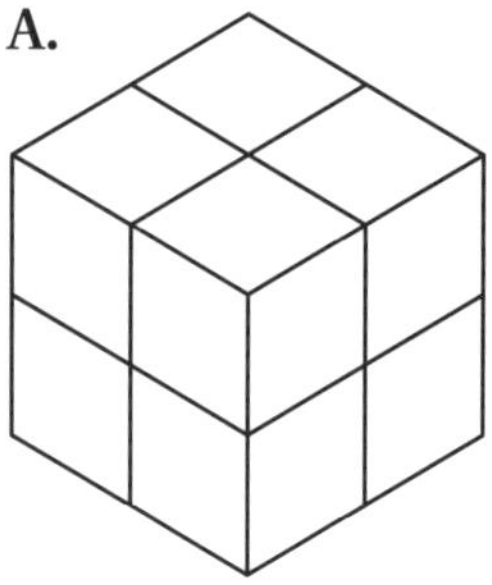

B.

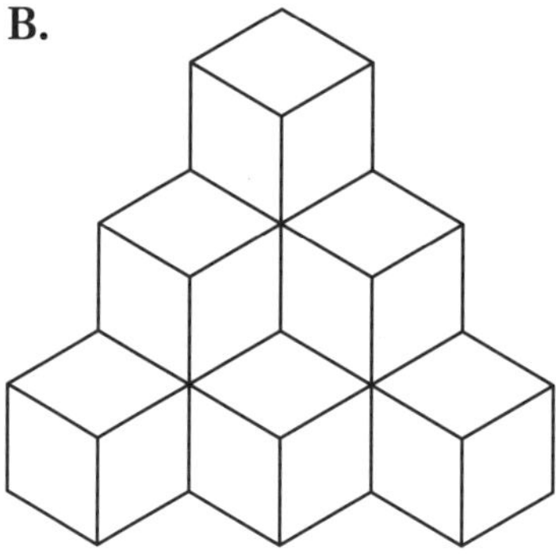

C.

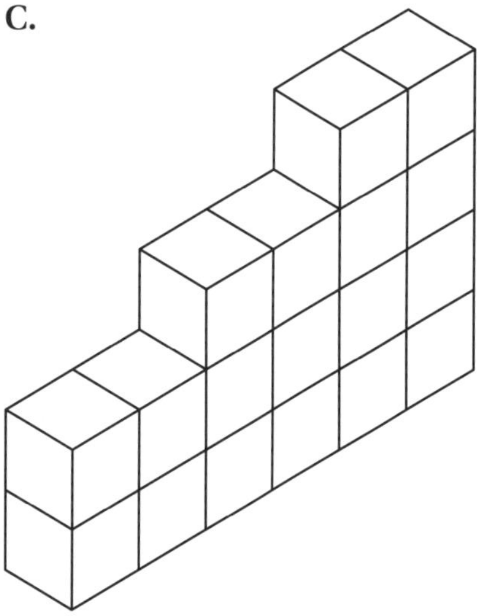

D.

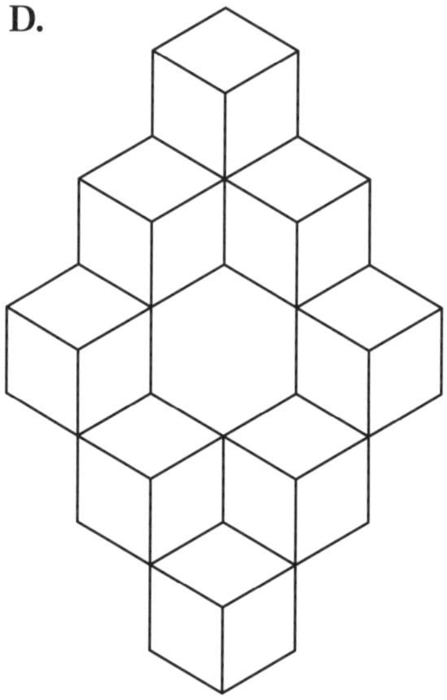

2. Which of these designs can be built as a stack of cubes? (The cubes can't be joined along edges.) What is the minimum number of blocks needed to build those stacks?
3. Build two other designs from cubes. Draw and shade them to create the illusion of three dimensions. Or draw designs that could be built from blocks and indicate how many blocks would be needed.

Nets of Cubes

1. Build the four cubes on Worksheet 13: Cut them out, fold on the dotted lines, and glue the flaps or tape the edges together. Describe what the nets have in common and how they are different.
2. Which of these drawings could be used as a net for a cube? For those that will not fold to a cube, explain why they won't.

A. 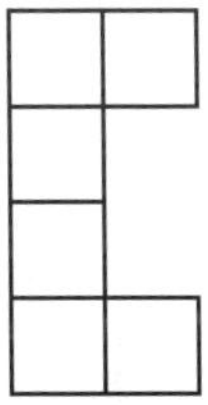

B. 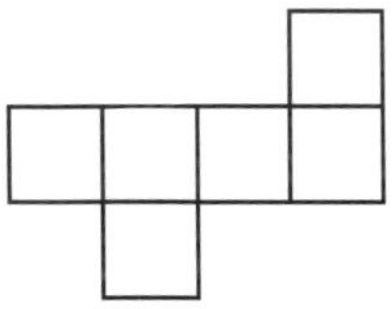

C. 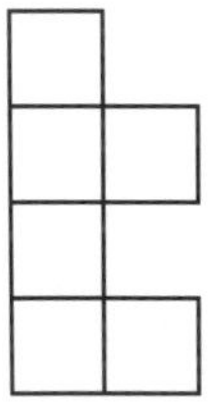

D. 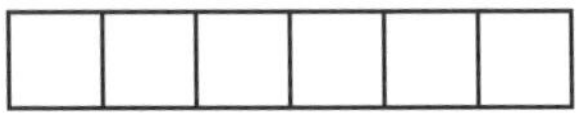

E. 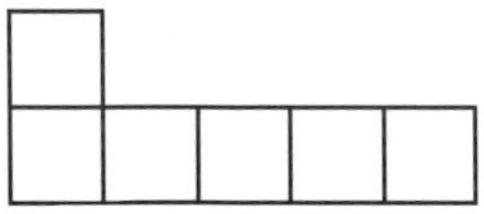

F. 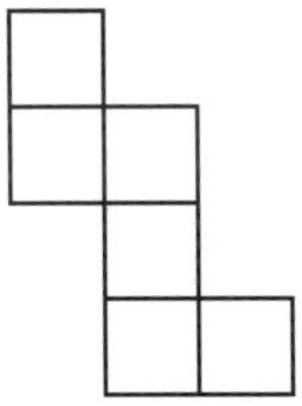

G. 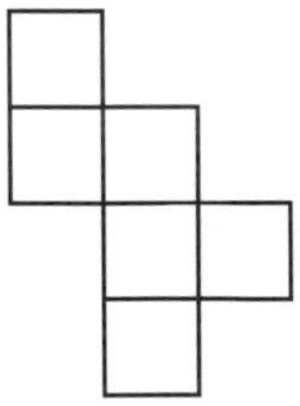

3. Draw a different net for a cube.

Nets of Cubes

WORKSHEET 13

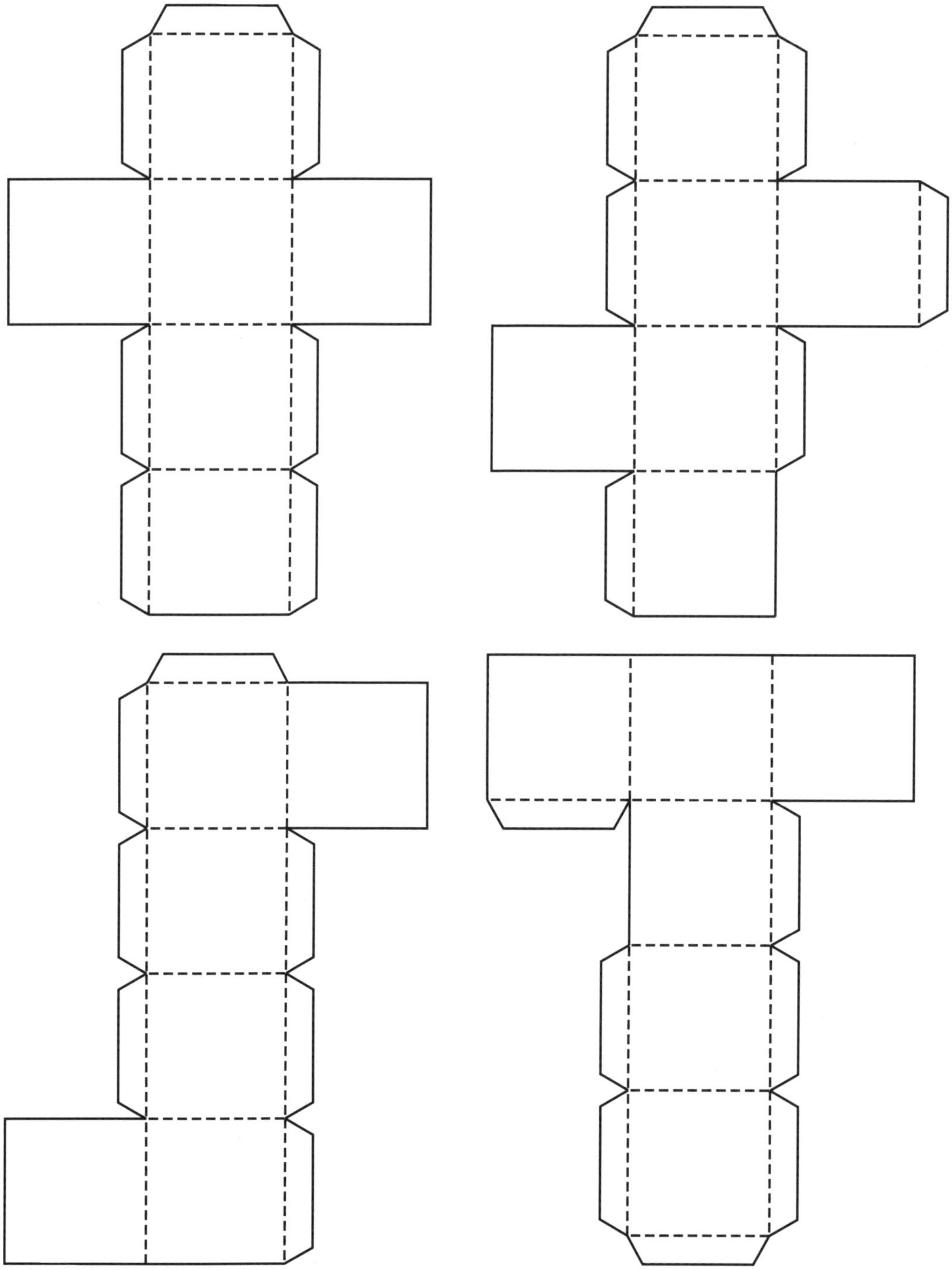

Perspective Drawing

Artists and geometers use *perspective lines, horizon lines,* and *vanishing points* to make two-dimensional drawings appear three-dimensional. For example, look at this drawing.

The telephone poles get smaller and smaller as they approach the horizon line, making them appear to recede into the distance. At the vanishing point, which lies on the horizon line, the objects seem to disappear from our sight.

Perspective lines follow the angle along the edges of a shape to the vanishing point. This sketch shows the perspective lines that allow our eyes to see the cube as a three-dimensional object.

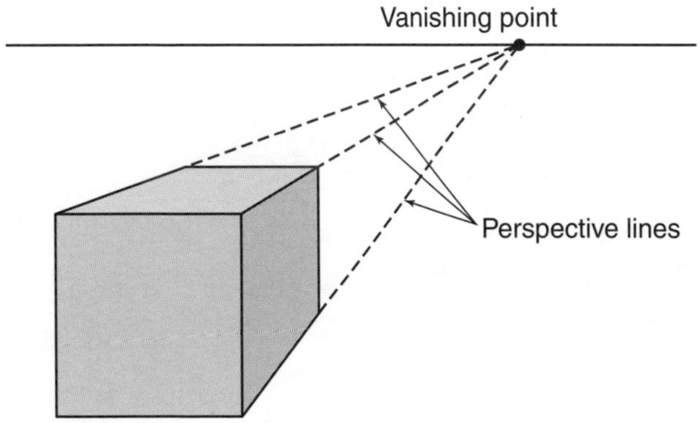

Choose a common three-dimensional shape, such as a rectangular box or a simple house, and sketch it using a horizon line, a vanishing point, and perspective lines. (Hint: Draw a horizon line and a vanishing point. Draw the two-dimensional front of the shape. Then draw dashed perspective lines from the corners of the front to the vanishing point. Use the perspective lines as a guide to complete the shape.)

ACTIVITY 4

Surface Area and Volume

The 20-triangle globe in the World of Geometry quilt is created from 20 equilateral triangles with $7\frac{1}{2}$ inches on a side.

1. Use the formula for the area of a triangle to calculate the approximate surface area of a globe made from 20 of these triangles.
2. Use your answer and the formula for the surface area of a sphere, $S = 4\pi r^2$, to calculate the radius of the globe.
3. Use the formula for the volume of a sphere, $V = \frac{4}{3}\pi r^3$, to calculate the volume of the globe.
4. What would be the dimensions for each triangle in a 50-triangle globe having the same surface area as the 20-triangle globe?

ACTIVITY 5

Images on a Cylinder

In this activity, you will observe what a design drawn on a semicircular wedge-shaped figure looks like projected onto a cylindrical surface.

For this activity, you will need Worksheet 14, colored pencils or markers, and a piece of reflective tailpipe tubing.

The worksheet contains a wedge-shaped grid and a rectangular grid, both marked with a coordinate system of numbers and letters.

1. Use colored pencils or markers to draw a design on the rectangular grid.
2. Transfer the design from the rectangle to the wedge, matching the corresponding coordinates. For example, if your design has a segment from A-8 to D-6, then draw a segment from A-8 to D-6 on the wedge grid.
3. Center the tailpipe tubing at the crosshairs marked on the worksheet. Compare the projection on the tubing with your drawing on the rectangular grid.

WORKSHEET 14

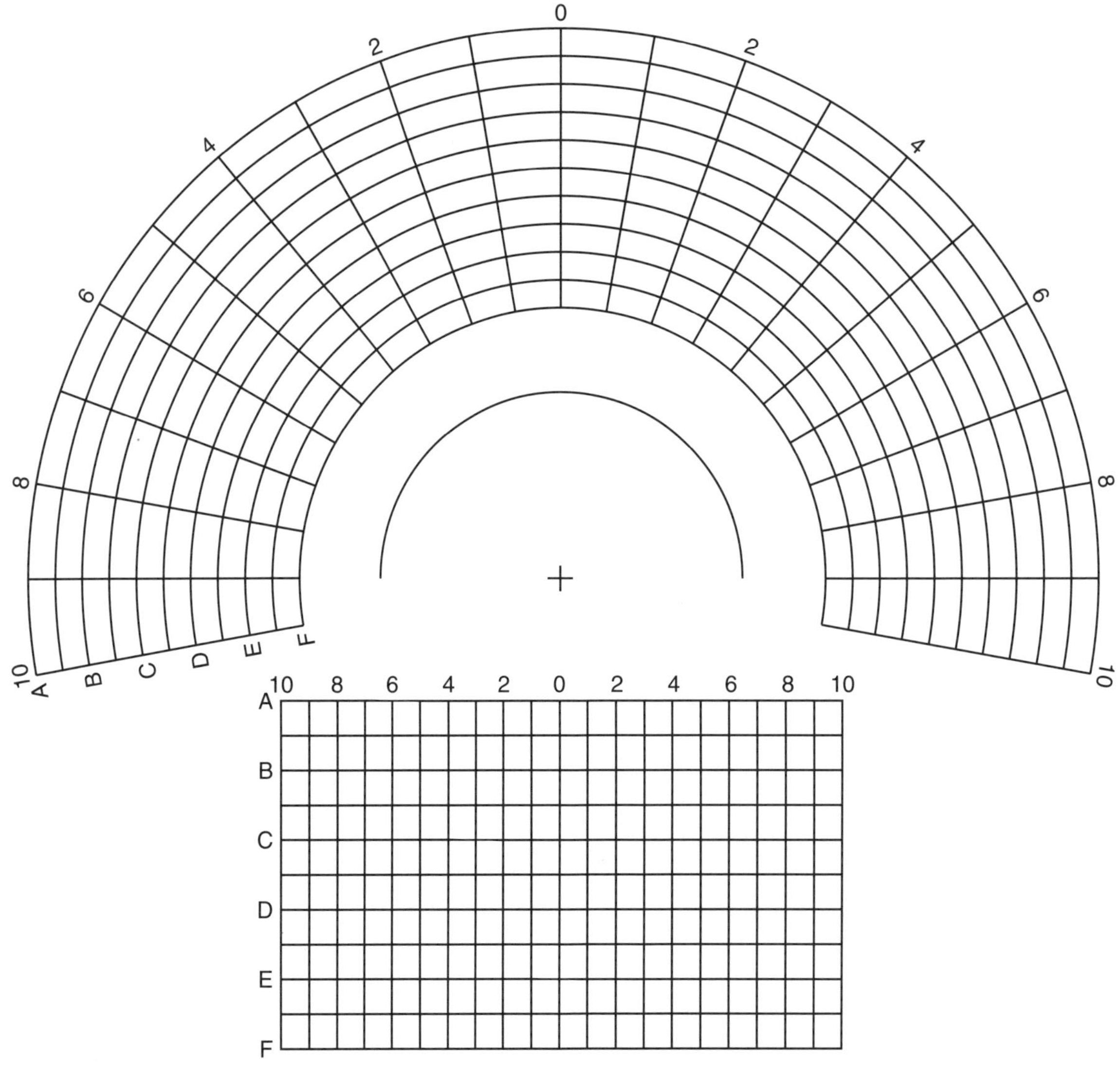

ACTIVITY 6

Flattening the Earth

Making a map of Earth requires representing the surface of a three-dimensional globe on a flat plane.

1. Draw some imaginary continents on the surface of a ball. Then cut up the ball and lay it flat. Describe what happens to the shape of the continents and to the distances and directions between parts of the ball. Explain why maps of smaller areas are less distorted than maps of larger areas.
2. Research to learn more about Mercator and gnomonic projections and others—including the polyconic, Mollweide, and sinusoidal projections. What is maintained in each projection, and what is distorted?

Mercator projection

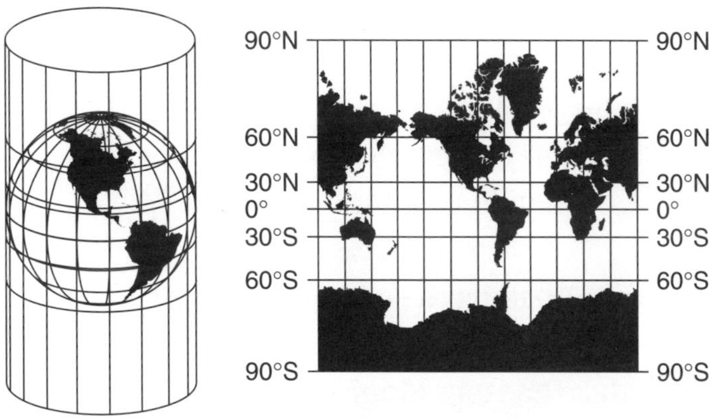

Gnomonic projection

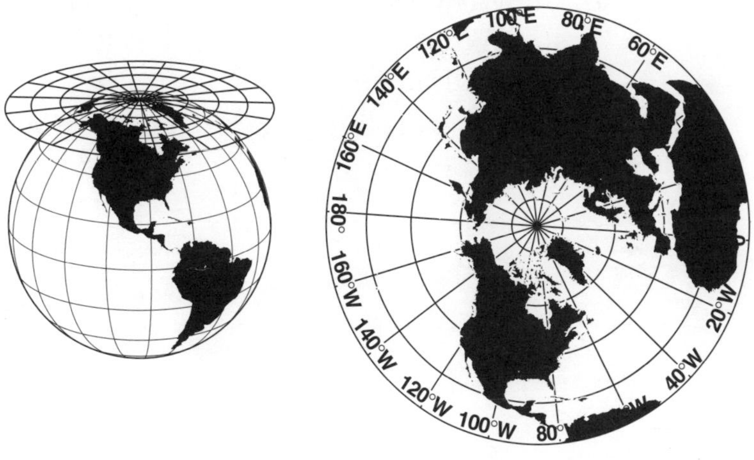

Möbius Strips

Topology is a branch of mathematics that studies surfaces and ways of distinguishing among surfaces; it studies the properties of geometric forms that remain unchanged when they are bent or stretched. One simple shape studied in topology is the Möbius strip, investigated in 1840 by the German mathematician August Möbius. The Möbius strip is created from a two-sided strip, but it has only one side.

To create your Möbius strip, you will need paper, a ruler, a pencil, scissors, and some tape. Cut a strip of paper about 11 inches long and 2 inches wide. Draw a dotted line lengthwise down the center of each side of the strip. Hold the two ends of the strip. Give the strip a half twist and tape the ends together. The front of one end of the strip should be taped to the back of the other end.

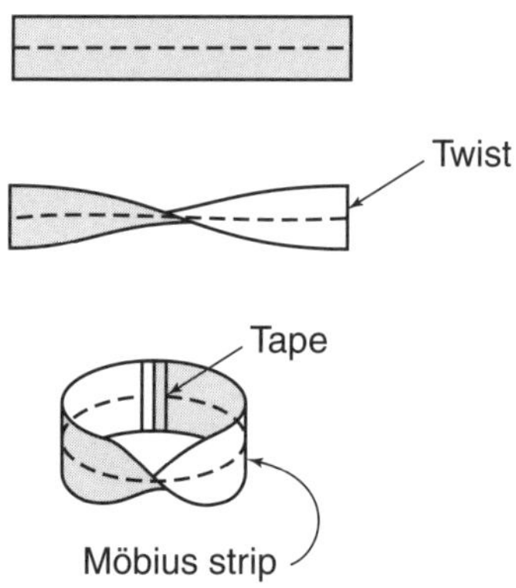

1. Using a pencil, trace over the dotted line on your Möbius strip. What do you discover?
2. Imagine cutting the strip along the line.
   a. What do you think the result would be?
   b. Cut along the line. What is the result? Why?
   c. Cut along the center of the result from 2b. What happens? Why?
3. Make another Möbius strip. (This time, do not draw lines down the center of the strip.)
   a. Draw a line about $\frac{1}{3}$ of the distance from the edge of the strip. What do you think the result would be if you cut along this line?
   b. Cut along the line. What is the result? Why?

Hypercubes

A *hypercube,* or *tesseract,* is an extension of the idea of a two-dimensional square and a three-dimensional cube extended to the fourth dimension.

1. Thinking about how a square is related to a segment and how a cube is related to a square can help you imagine how a hypercube is related to a cube. Fill in the table with the dimensions and number of vertices and edges of a point, segment, square, cube, and hypercube. Describe the pattern.

	Point	Segment	Square	Cube	Hypercube
Dimensions	0	1			4
Vertices	1				
Edges	0				

2. The shadow of a cube can be drawn in several ways.

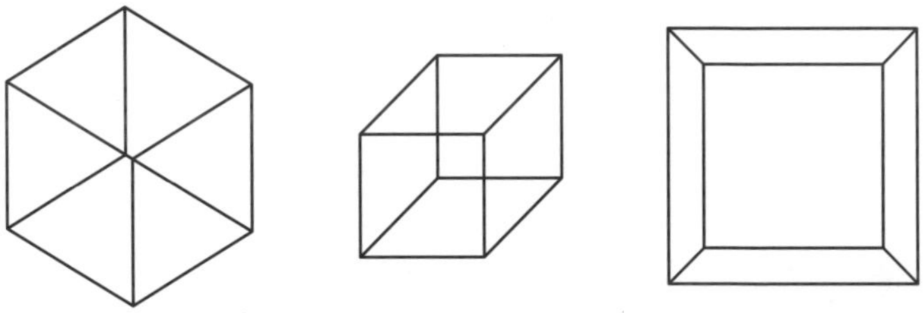

The first way is looking down on a cube; the side lengths are distorted, but no lines overlap. The second way connects two squares that are the same size. The third way is looking at one vertex and seeing the diagonally opposite vertex right next to it. The three-dimensional shadow of a hypercube can also be built in these three ways. Describe one of these ways and build the model.

ACTIVITY 9

Bubble Models

For this activity, you will need a container of bubble solution, two pieces of string, and something from which to build edges and vertices of polyhedra.

Create a model of the vertices and edges of a cube and of a tetrahedron.

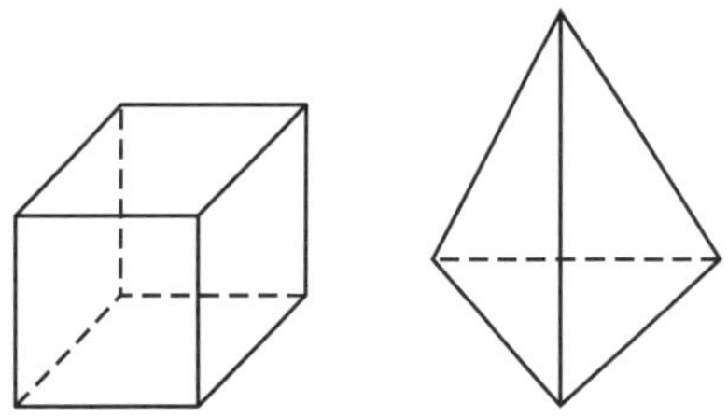

To build the models, you can use

Toothpicks or wire for the edges and modeling clay for the vertices

Straws held together by string

A building set, such as the Zome System

Tie a string to each model. Hold the cube by its string and dip it into the bubble solution. Lift the model out of the solution. If all the vertices are connected with bubble film and a small cube appears within the larger cube, you have created a projection or shadow of the hypercube.

You can improve the projection by dipping the straw into the bubble solution and then inserting it into the center cube of the hypercube shadow. Blow gently and the center cube will expand. You are looking at a three-dimensional model of a four-dimensional hypercube's shadow.

A four-dimensional tetrahedron is called a *simplex*. Dip the model of the tetrahedron into the bubble solution to create a projection of a simplex. Describe what you see.

ACTIVITY 10

Research Topics

1. Research the life of one of these mathematicians, all of whom investigated non-Euclidean geometry. Be sure to discuss their contributions to non-Euclidean geometry.

   Bolyai, Janos
   Gauss, Carl Friedrich
   Lambert, Johann
   Lobachevsky, Nikolai
   Riemann, Bernhard
   Saccheri, Girolamo

2. In four-dimensional space, you could escape from the inside of a sphere without going through its surface! Or you could turn a sphere inside out without tearing it. Here are some other interesting facts about what life would be like in a four-dimensional world.

   You could enter or leave a building without using doors and windows.

   A surgeon could operate on a heart without injuring the skin or the walls of the heart.

   A knot could be untied without moving the ends of the rope.

   The contents of an egg could be removed without breaking the shell.

   If you are intrigued by these ideas, read and report on the article "Surreal Films: A Soapy Solution to the Math Puzzle of Turning a Sphere Inside Out" in *Science News,* October 10, 1998.

3. Below are some additional topics to research.

   Banchoff, Thomas
   cartography
   Clifford, William
   Descartes, René
   Einstein, Albert
   fourth dimension
   Fuller, Buckminster
   geodesic domes
   hypercube
   hypersphere
   Lénárt sphere
   Leonardo da Vinci
   Möbius, August
   perspective
   third dimension
   topology
   Young, Grace Chisholm

Teacher Notes

The design in the Tumbling Blocks Pyramid quilt can be created on isometric dot paper or on Worksheet 11. You might ask what is the smallest number of cubes needed to build the tumbling blocks pyramid so that it will not tumble. (There would be at least 84 cubes in the stack. The 84 cubes would be stacked in a triangular pyramid with right triangles as the two faces that are not visible.) Activity 1 gives students practice making isometric drawings. Allowing students to build models of the figures from cubes will give them experience in noticing which faces of the three-dimensional object are missing in the drawing and which cubes are not visible at all. Activity 2 helps improve students' spatial visualization skills as they imagine nets folded into cubes and also their dexterity as they cut and fold the cubes. In Activity 3, students look at *perspective* drawing, another way of representing three-dimensional objects in two dimensions.

Building a paper model from the icosahedral or from the 50-triangle patterns for the World of Geometry quilt is an ideal group project. Students could build other polyhedra and discuss how well they would work as models of a globe. Once the models are built, students could research historical information about the polyhedra. Many of the polyhedra will have connections to chemical structures. In Activity 4, students use the formulas for the area of a triangle and the surface area of a sphere to help them calculate the volume of the 20-triangle globe.

Activity 5 gives students an idea of how one shape can change into another shape by means of a projection. In Activity 6, students explore what happens when a sphere is flattened and learn about projections used to create maps. Extend these activities by discussing or by having students research the history of mapmaking and modern mapmaking methods. Your students might be interested in learning more about the Global Positioning System (GPS)—a system of satellites used to determine exact locations on Earth.

The Clifford torus gets students thinking about other dimensions. The Möbius strip, which students explore in Activity 7, is a fun model of a one-dimensional space that has no boundary. To help them imagine a two-dimensional universe, students could read *Flatland* by Edwin Abbott. This satire of Victorian society in a flat world is about 100 pages long. Another interesting book is *Sphereland* by Dionys Burger.

In Activities 8 and 9, students describe and create models of four-dimensional figures. Your advanced students might be interested in reading *Beyond the Third Dimension—Geometry, Computer Graphics, and Higher Dimensions* by Thomas F. Banchoff. It includes examples of CAT scans, MRI images, X-ray images, flower heads, and contours. It also discusses research into the dimensionality of rehabilitation therapy, pendulum patterns, and paleoecology (a branch of ecology concerned with the characteristics of ancient environments and their relationships to ancient plants and animals).

Activity 10 gives students the opportunity to expand their knowledge of related fields in mathematics. Many of the topics listed here can help instill further interest in contemporary mathematics topics not covered in the traditional classroom.

Answers and Comments

Activity 1

1. Drawings should match the art shown on the activity page.
2. Eight blocks are needed to build A, 10 blocks for B, 18 blocks for C, and D cannot be built.
3. Designs will vary.

Activity 2

1. The nets each have six faces with four in a row. The two faces not in a row are attached to different sides of the strip of four. The only difference is where the two side pieces are attached.
2. A and C won't work because there would be two squares on top of each other on one side of the box and no square for the opposite face. D and E won't work because there would be five or six pieces around the sides, but there can only be four. Nets B, F, and G are nets for cubes.
3. Nets for cubes include:

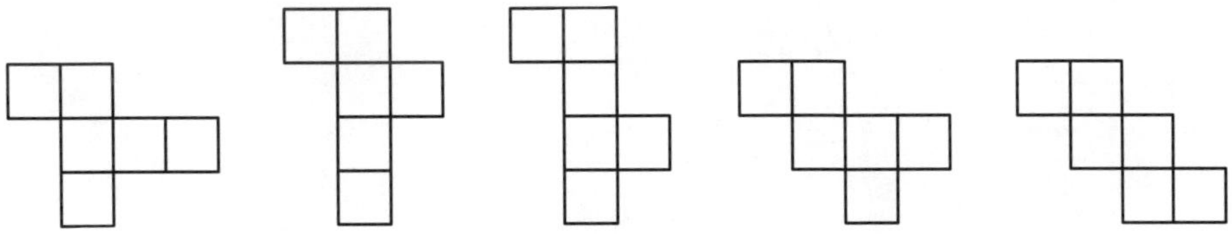

Activity 3

Perspective drawing can increase students' observational skills. Look for individual skill and improvement.

Activity 4

1. The base of the triangle is 7.5 inches, and the height is about 6.5 inches, so the area is about 24.4 square inches. Because the globe is made from 20 triangles, its surface area is $20 \cdot 24.4$, or 1218 square inches.
2. $4\pi r^2 = 1218$ square inches, so $r \approx 9.8$ inches.
3. $V = \frac{4}{3}\pi(9.8)^3 \approx 3960$ cubic inches
4. The globe is the same size, so the surface area will also be 1218 square inches. Divide by 50; each triangle will have an area of about 2.44 square inches; if the triangles are equilateral, the length of each side will be about 2.4 inches.

Activity 5

For this activity, each student will need a piece of reflective tailpipe tubing about 6 inches long and 2 inches in diameter. The tubing is available at most hardware stores for a reasonable price. If a tailpipe is not available, a toilet paper roll covered with Mylar or other reflective material can be used to view the projection. A fun resource with activities similar to this one is *The Magic Cylinder Book* by Ivan Moscovich.

Sample answer:

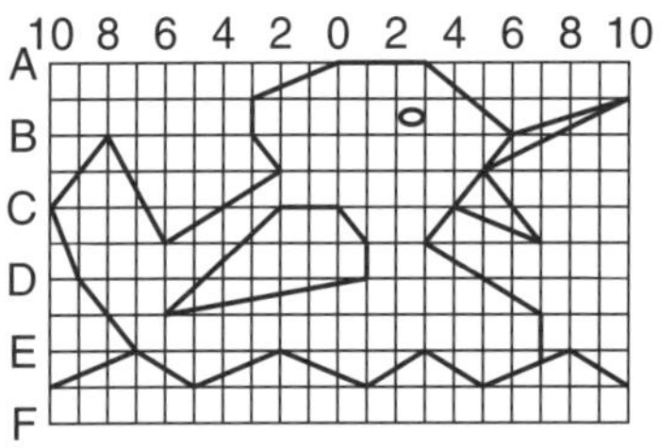

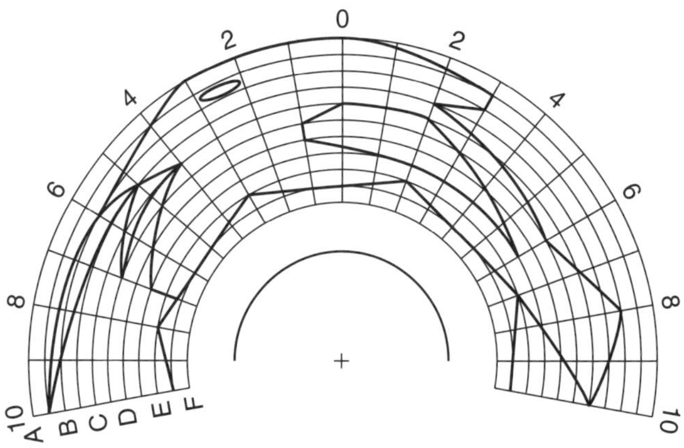

Activity 6

1. Sample answer: The shapes of some of the continents were distorted. Some continents were cut and spread out. The distances between points on the opposite sides of a cut increased when the ball was flattened. Directions between points also changed.

   Another sample answer: If you think of Earth as a ball, then making a map of a small area is like cutting a small piece from the ball and flattening it. Flattening a small area requires less cutting; it causes less spreading than flattening a large area. Therefore, the map will be less distorted.

2. Answers may include: The *Mercator projection* is created by projecting Earth onto a cylinder tangent to Earth at the equator. This projection shows Earth, except the extreme polar regions, on one continuous map. True directions such as north and south are shown as straight lines, but distances at high and low altitudes are distorted.

   Projecting Earth onto a plane tangent to one of the poles creates a *gnomonic projection*. This projection correctly shows the shortest route between two points, but it distorts directions and distances.

Activity 7

1. The lines that were on different sides of the original strip now form one continuous line. The Möbius strip has only one side (surface) even though it is constructed from a two-sided strip of paper.

2. a. Answers will vary.
   b. Instead of coming apart to form two loops, the strip becomes one large loop with two full twists. Possible explanations might include the idea that the cut makes a second edge, so now the strip has two sides.
   c. After the second cut, there are two interlocking loops, each with two full twists. Possible explanations might include the idea that the cut makes two more edges, so there are now four edges and four sides.

3. a. Answers will vary.
   b. Two interlocking loops are created; one is a Möbius strip, and the other is a one-sided strip with three half twists. Possible explanations might include the idea that the one cut went around the surface twice.

Activity 8

A three-day version of this activity is presented in the October 1995 issue of *Mathematics Teacher*. The article has the background for the teacher, step-by-step procedures to assemble a 104-straw hypercube, worksheets, and answers.

1.

	Point	Segment	Square	Cube	Hypercube
Dimensions	0	1	2	3	4
Vertices	1	2	4	8	16
Edges	0	1	4	12	32

Patterns include: As the dimension increases by one, the number of vertices doubles. The number of edges in n dimensions is equal to twice the number of edges plus the number of vertices in $n - 1$ dimensions ($32 = 2 \cdot 12 + 8$; $12 = 2 \cdot 4 + 4$; $4 = 2 \cdot 1 + 2$; $1 = 2 \cdot 0 + 1$).

2. Start with two cubes of different sizes, one inside the other, and connect them. Start with two cubes of the same size and connect their vertices.

   Start with a rhombic dodecahedron, add two new vertices in the center, and connect each to four of the vertices (for a complete description, see pages 175–177 in *Zome Geometry* by George W. Hart and Henri Picciotto).

Activity 9

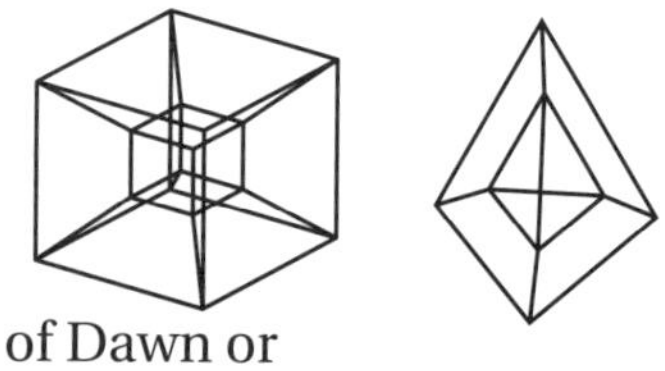

The bubble will form a small cube or tetrahedron within the wire frame.

For bubble solution, you can use a commercial solution like Wonder Bubbles. Or use this recipe from the Exploratorium in San Francisco: To each gallon (3.8 liters) of water, add $\frac{2}{3}$ cup (160 ml) of Dawn or other dishwashing liquid and 1 tablespoon (15 ml) of glycerine, available at your local pharmacy. The bubble solution works best if it is aged at least a day before use. The glycerine helps bubbles last longer, but it's not essential.

Even More Mathematical Quilts

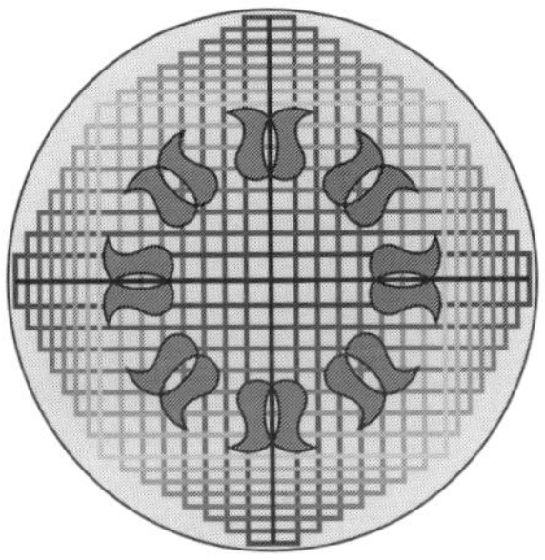

The Leonardo's Dessert Quilt

The Pascal's Surprise Quilt

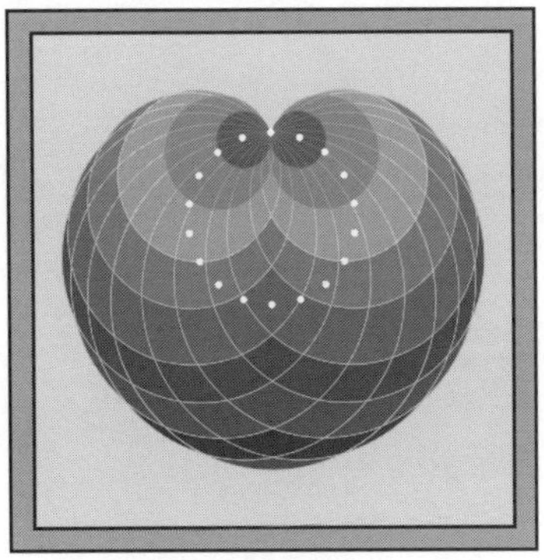

The Mascheroni Cardioid Quilt

THREE MATHEMATICIANS, WHOSE LIVES SPAN 350 years of history, inspired the three quilts in this section. Each had a different area of interest. Leonardo da Vinci (1452–1519) explored geometry, Blaise Pascal (1623–1662) was intrigued by probability, and Lorenzo Mascheroni (1750–1800) studied compass constructions.

Leonardo da Vinci, the famous Italian artist who painted the *Mona Lisa* and the *Last Supper*, was also a sculptor, inventor, civil and military engineer, botanist, astronomer, geologist, and geometer. His notebooks and sketches are still studied by scholars today. Leonardo's ideas were so far ahead of his time that many of them did not become realized until the twentieth century. He is considered one of the greatest thinkers in history.

The design used in the Leonardo's Dessert—Sherbet Without Pi quilt is based on Leonardo's work with curved shapes. He was interested in seeing which curved shapes could be cut in such a way that the pieces could be reassembled to form a rectangle. For example, the drawing below shows how a "clamshell" shape can be made into a rectangle with the same area.

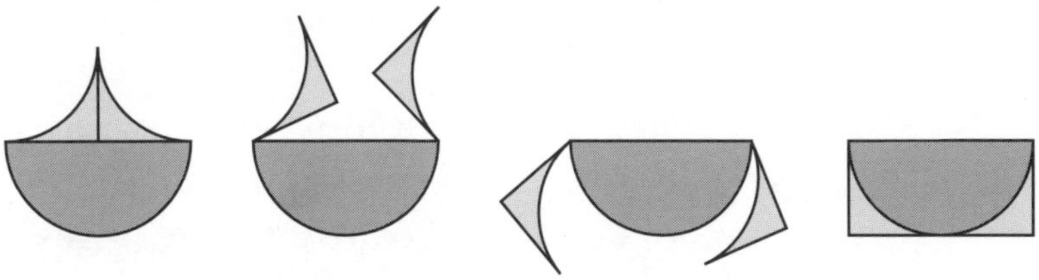

The "hinged dissection" shape in the quilt is derived from the clamshell.

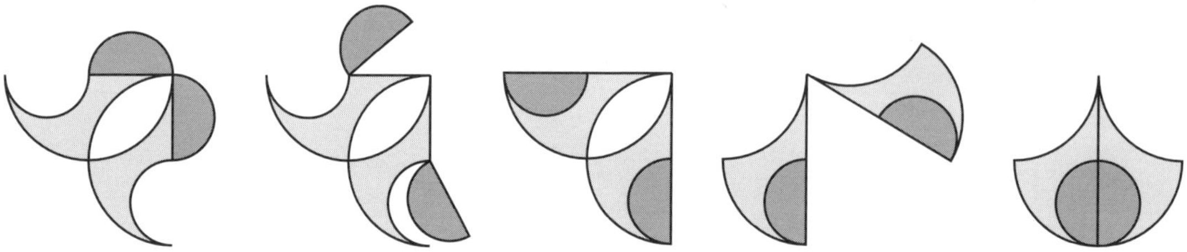

Leonardo's drawings of these curved shapes could be the result of his experiments to construct a square with the same area as a given circle. This process, called *squaring the circle,* is now known to be impossible, but it was of great interest to mathematicians in Leonardo's time.

Pascal's triangle was named for Blaise Pascal, who explored this triangular array of numbers, which was known and studied long before his time. There is evidence that the Persian poet, astronomer, and mathematician Omar Khayyám (1050–1123) worked with the triangle, and the Chinese algebraist Chu Shih-chieh (1280–1303) used the number pattern. The pattern used to create Pascal's triangle of numbers is based on addition: Each row begins and ends with 1, and each of the other numbers is the sum of the two numbers diagonally above it.

1
1 1
1 2 1
1 3 3 1
1 4 6 4 1
1 5 10 10 5 1
1 6 15 20 15 6 1
1 7 21 35 35 21 7 1
1 8 28 56 70 56 28 8 1
1 9 36 84 126 126 84 36 9 1
1 10 45 120 210 252 210 120 45 10 1

Pascal's triangle contains many patterns, including the one used for the quilt, a pattern explored by mathematics teacher Mark Roth. The details of Roth's design are given on pages 138–140. Pascal's triangle is related to many mathematical topics, including probability, combinatorics, algebra, and fractal geometry.

Many ideas are named for the person who introduces it to the world. But as with Pascal's triangle, the person whose name is given to a mathematical concept may not be the first person to discover or use it. Although compass-only constructions are known as Mascheroni constructions, Danish mathematician Georg Mohr (1640–1697), not Lorenzo Mascheroni, was the first to show that all compass-and-straightedge constructions could be accomplished with a moveable compass alone. (See "A History of Compass Constructions.") The design in the Mascheroni Cardioid quilt was given its name because the cardioid it outlines was constructed using only a compass.

The outline of a cardioid can be created by drawing a circle and then constructing other circles centered at evenly spaced points on the original circle, all going through a common point on the first circle. A *cardioid*, meaning "heart-shaped," is the locus (path) of a point on a circle rolling around a fixed circle of the same diameter. It can also be thought of as the graph of the polar equation $r = 2a\,(1 \pm \cos\theta)$. The Greeks investigated cardioids to describe the motion of the planets as "circles moving upon circles." Danish astronomer Ole Rømer (1644–1710), who was seeking the best shape for gears, studied the cardioid, along with related curves, in the late seventeenth century.

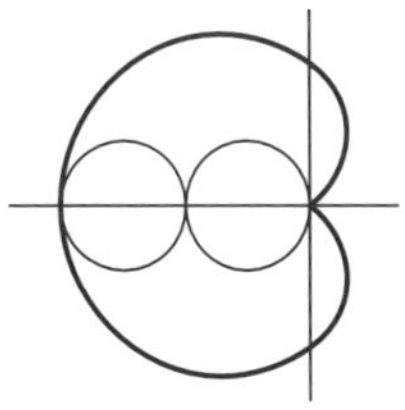

The History of Compass Constructions

The task of geometric construction goes back to the ancient Egyptians, but the Greek philosopher Plato (ca. 427–347 BCE) set the current standards. Although the Greeks had many instruments for constructing figures, including an angle trisector, constructions were limited to an unmarked straightedge and a compass because Plato believed that all motion was either along straight lines (on Earth) or circles (in the heavens). The straightedge can be used only to draw a line segment, not to measure; and the compass can be used only to draw a circle with a given center and radius. In Plato's time, the compass device used was not portable, so transferring distances was quite complex. When a compass was raised from the sand tray in the drawing that was made, it collapsed.

Euclid (ca. 300 BCE), a mathematician who lived and worked in Alexandria when it was the center of Greek learning, compiled most of the mathematics known in his time into *The Elements*. One of the most important and influential mathematics texts ever written, it includes extensive coverage of compass-and-straightedge constructions, including a proof that a line could be copied into a different place by using a collapsing compass. Thus, a noncollapsing compass could be used to transfer a distance, with the understanding that the process could be achieved more laboriously with a collapsing compass. The tenth-century Persian mathematician Abul Wefa (940–998) described constructions that were possible with a straightedge and a *fixed compass*, that is, a compass whose radius cannot be changed. The moveable compass of today came into use after the tenth century.

Around 1484, the German artist Albrecht Dürer (1471–1528) described how a regular hexagon could be constructed using a fixed compass. Dürer's interest in the construction of regular polygons stems from his interest in Gothic and Islamic decorative arts, which contain many polygons. His most important book, *Investigation of the Measurement with Circles and Straight Line of Plane and Solid Figures*, contains some striking novelties and curves.

In 1673, Georg Mohr published *Euclides danicus*, in which he proved that every compass-and-straightedge construction can be done with a moveable compass alone. Of course, a straight line cannot be drawn with a compass, so in compass-only constructions, straight lines through points must be imagined.

Many Renaissance mathematicians, including Leonardo da Vinci, experimented with fixed-compass constructions. In the nineteenth century, the French mathematician Jean Victor Poncelet (1788–1867) and the Swiss mathematician Jakob Steiner (1796–1863) developed mathematical proofs that all compass-and-straightedge constructions can be done with a straightedge and fixed compass. In 1797, Lorenzo Mascheroni, an Italian mathematician, published *Geometria del compasso*. Just as Mohr had done 125 years earlier, Mascheroni proved that all compass-and-straightedge constructions could be done using a moveable compass alone. Because Mohr's work had gone virtually unnoticed at the time it was published, it was Mascheroni who was given credit for this result. In fact, compass-only constructions are often referred to as *Mascheroni constructions*. Mohr was not given credit for his discovery until 1928, when a copy of his book was found in a Copenhagen bookstore.

The Leonardo's Dessert—Sherbet Without Pi Quilt

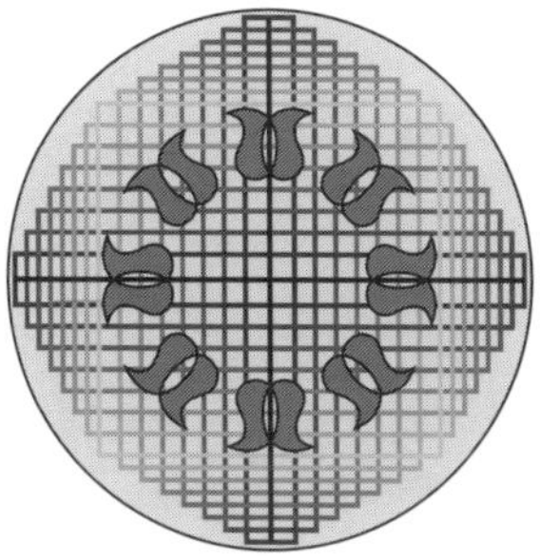

The curved shape repeated in this quilt was found in the *Codex Atlanticas,* a collection of Leonardo da Vinci's notes and figures. It is one of the many shapes he studied that can be cut into pieces and rearranged to form a rectangle.

The quilt was named Leonardo's Dessert—Sherbet Without Pi because geometry was an intellectual dessert for Leonardo. And in his work, Leonardo used only intuitive geometry rather than doing calculations involving the number pi. The word *sherbet* in the title refers to the colors of the quilt background.

Designing Your Quilt

1. The steps here describe how to make one of the curved figures in Leonardo's Dessert quilt. To make the quilt design, you need eight copies of this shape. You can draw eight shapes separately or make one and photocopy it.

Step 1 Start with a square and draw a horizontal line segment dividing the square in half. Cut on the line and use just half of the square.

Step 2 Locate the midpoint of the long edge of the half square. Put the sharp end of your compass on this point and adjust the opening so the pencil end touches the point where the horizontal segment meets the square. Then swing the compass to draw a semicircle in the half square.

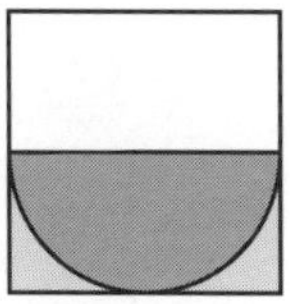

Step 3 Cut off the two corners along the semicircle lines.

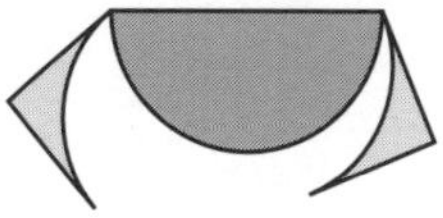

Step 4 Rotate the cutoff pieces as shown so the straight sides align with the diameter of the semicircle, and tape the pieces in place. This shape is called a *clamshell.*

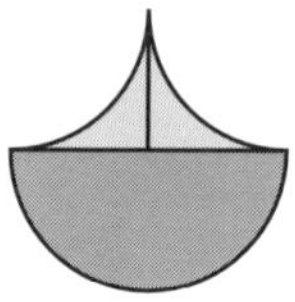

Step 5 Draw a vertical line segment that bisects the clamshell. Mark a point on this segment $\frac{1}{4}$ of the distance from the bottom of the clamshell to the top. Place the sharp end of your compass on this point. Adjust the opening so the pencil point touches the bottom of the clamshell and draw a circle.

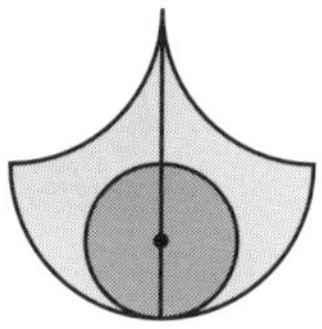

Step 6 Cut the clamshell along the vertical bisector, and rotate the right half counterclockwise around the top vertex until it meets the left half. Tape the pieces in place.

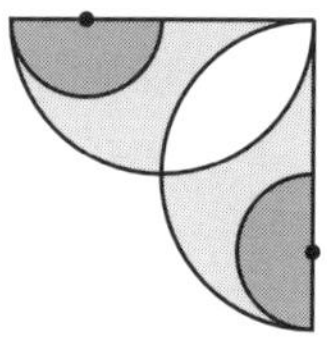

Step 7 Cut out the two small semicircles, rotate them as shown, and tape them in place.

The result is the shape that appears in the quilt. Look at the shape and remind yourself that it began with a half square. Like Leonardo da Vinci, you have created a curved figure, often called a *hinged dissection*, from a rectangle.

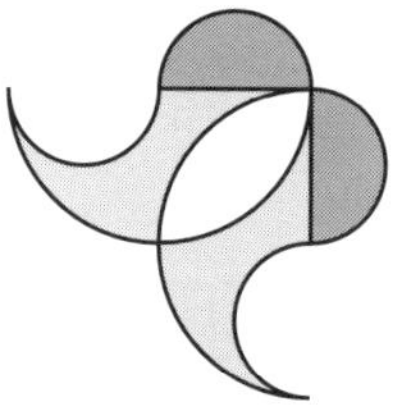

You may want to erase or cover the construction lines in the shape.

2. Arrange eight copies of the shape in a circular pattern. To help you place the figures, you can draw a light circle. Dividing the circle into fourths or eighths can help you space the shapes evenly. When you are finished arranging the shapes, draw an interesting background pattern for your design and add color.

Making the Leonardo's Dessert—Sherbet Without Pi Quilt

The quilt is a 45-inch diameter circle; the background design was created using $\frac{1}{2}$-inch strips of a variety of colors. The gold lamé hinged dissection figures are 9 inches wide and 8 inches long. If you follow the instructions on pages 133–134 beginning with a 7-inch square, you will generate the 9-by-8-inch figure.

Materials

$2\frac{1}{2}$ yards of white background fabric

$\frac{1}{2}$ yard each of fabric or bias tape in three colors (green, orange, and yellow) for the background design

$1\frac{1}{2}$ yards of gold lamé fabric

Gold lamé quilting thread

2 yards of gold cording

Batting (50 by 50 inches)

Instructions

Draw the design of the quilt on freezer paper to help you visualize the gold hinged dissection shapes against the background.

Cut the background fabric into a $45\frac{5}{8}$-inch diameter circle (the $\frac{5}{8}$ inch is a seam allowance for adding the border).

To position the fabric strips, mark the circumference of the circle with 64 equal arcs. Align the strips, using the marks as a guide, parallel to one another and symmetrically about the center of the circle. Interweave the fabric strips.

Next, create the 8 hinged dissections by tracing the 16 copies of the half shapes and 16 copies of the mirror image of the half shape onto the material.

Cut out two mirror image shapes for each half of a hinged dissection, adding a $\frac{1}{2}$-inch seam allowance. Place the right sides together and stitch $\frac{1}{2}$ inch from the edge. Stitch nearly all the way around, turn, and stitch the remaining gap by hand. If you are using lamé, it is important not to iron it. Place 16 pieces in 8 figures around the circumference of a circle on the lattice. Each half shape should be appliquéd separately.

Layer top, batting, and backing. Quilt to outline the shapes with gold lamé thread. Add other quilting lines as desired. Add gold piping to the border of the quilt.

To hang a circular quilt, make a fiberboard circular frame, glue Velcro to one side of the fiberboard, and sew the opposing Velcro along the circumference of the quilt back. Stick the quilt to the frame and hang the frame.

The Pascal's Surprise Quilt

Blaise Pascal, a seventeenth-century French mathematician and scientist, made many contributions to physics, probability, and geometry. His name is often associated with a fascinating triangle of numbers he used while studying probability, known today as *Pascal's triangle.*

1
1 1
1 2 1
1 3 3 1
1 4 6 4 1
1 5 10 10 5 1
1 6 15 20 15 6 1
1 7 21 35 35 21 7 1
1 8 28 56 70 56 28 8 1
1 9 36 84 126 126 84 36 9 1

Each row of Pascal's triangle begins and ends with the number 1; each of the other numbers is the sum of the two numbers diagonally above it.

His pattern involves writing the units digit of the Pascal's triangle numbers in upward-pointing triangular cells of a triangular grid. The three numbers surrounding each empty grid cell are added, and the units digit of the sum is written in the cell. Then colors are assigned to each cell, depending on the number in the cell.

The Pascal's Surprise quilt is based on a design by Marc Roth, a mathematics teacher in California. To create his design, he first replaced the numbers in Pascal's triangle with their mod-10 equivalents. In *mod 10,* each number is equivalent to the remainder when 10 divides the number. For example,

$$25 \div 10 = 2 \text{ R5, so } 25 \text{ mod } 10 = 5$$

$$57 \div 10 = 5 \text{ R7, so } 57 \text{ mod } 10 = 7$$

$$3 \div 10 = 0 \text{ R3, so } 3 \text{ mod } 10 = 3$$

$$300 \div 10 = 30 \text{ R0, so } 300 \text{ mod } 10 = 0$$

As you can see, the mod-10 equivalent of a number is simply the units digit of the number. Here is Pascal's triangle with the numbers written in mod 10 on a triangular grid.

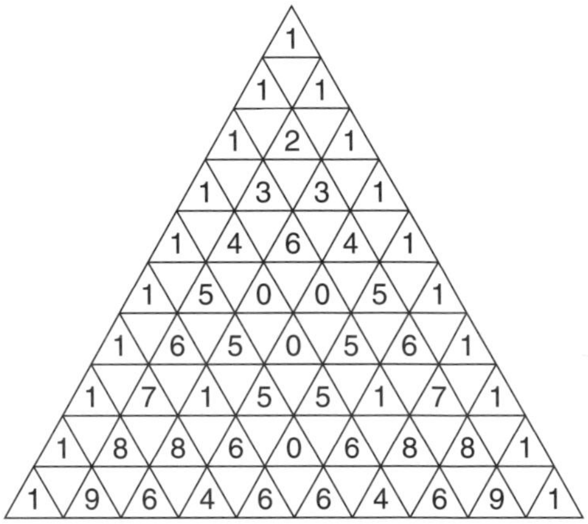

Roth filled in each empty cell with the sum of the three numbers that surround it, writing the sum in mod 10. Here are the first few rows.

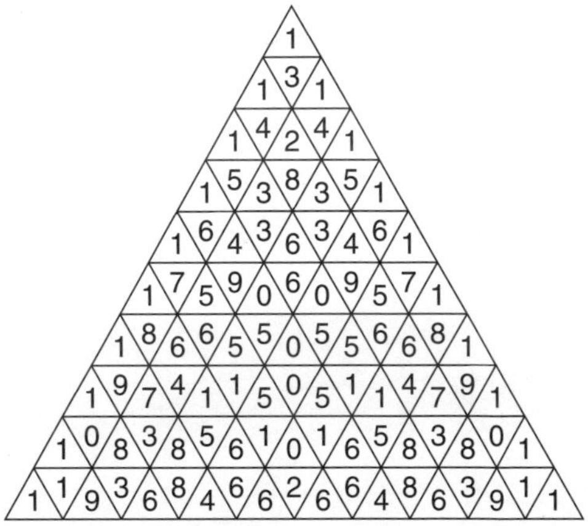

He then shaded all the cells with odd numbers black, and used red, yellow, and blue to shade the cells with even numbers.

Designing Your Quilt

To create your Pascal's Surprise design, you will need Worksheet 16, page 141, and markers or colored pencils in four colors.

Step 1 On the triangular grid, reconstruct Pascal's triangle in the upward-pointing triangles so that all the numbers are in mod 10. In other words, show only the ones digit. Use a pencil and make the numbers light so they will not show through your colors.

Step 2 For each blank cell in your reconstructed triangle, add the three numbers surrounding the cell, convert the sum to mod 10, and write the result in the cell.

Step 3 Use one color to shade all the cells containing odd numbers. On the quilt, black is used.

Step 4 Look for groups of four unshaded cells that form triangles. Use a second color to shade *all* four-cell triangles that contain even numbers.

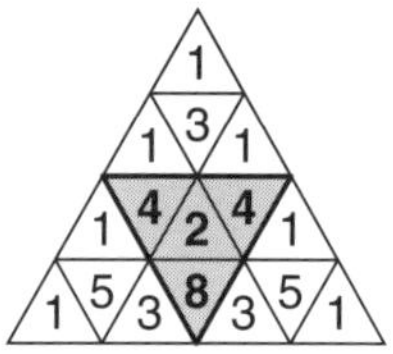

Step 5 Look for pairs of unshaded cells that form parallelograms. Use a third color to shade these two-cell parallelograms that contain even numbers.

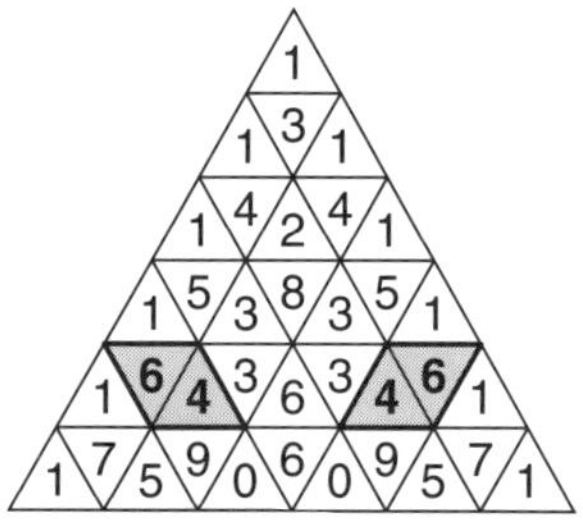

Step 6 Use the fourth color to shade the remaining cells.

Triangular Grid for Pascal's Surprise Quilt Design

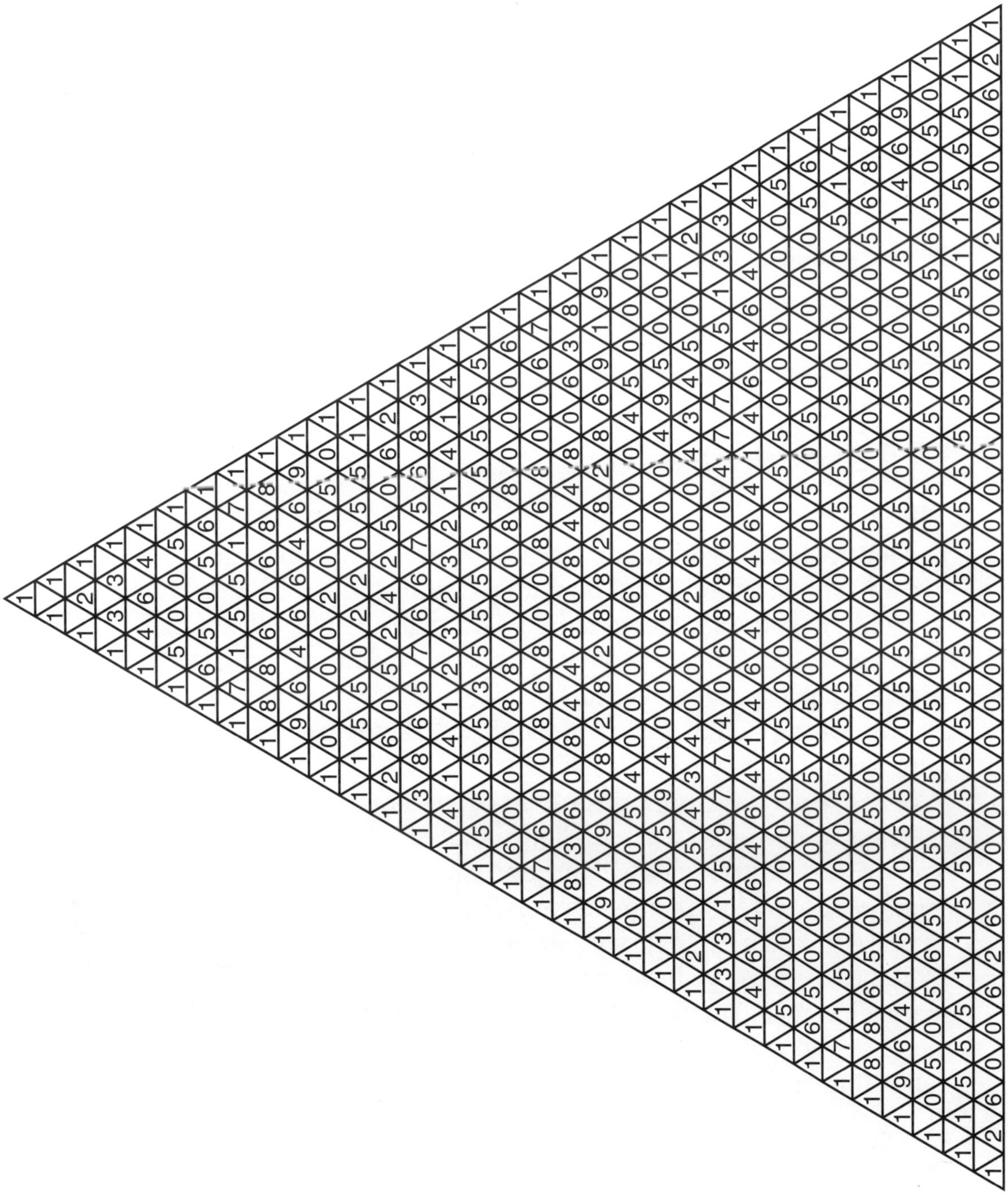

Making the Pascal's Surprise Quilt

The Pascal's Surprise quilt measures 50 inches by 42 inches.

Materials

$\frac{3}{4}$ yard of yellow fabric

$\frac{1}{2}$ yard of red fabric

$\frac{3}{4}$ yard of blue fabric

$2\frac{1}{2}$ yards of black background fabric

1 yard of 45-inch border fabric

$1\frac{1}{2}$ yards of 45-inch fabric for backing

45-by-60-inch batting

Freezer paper

Instructions

The entire design is drawn on freezer paper as a guide to the size and shape of the quilt. The quilt is composed of equilateral triangles, with the smallest triangle having an altitude of 1 inch. To prevent stretching, use freezer paper as a stabilizer. As you work, cut each piece to the exact finished size from freezer paper and iron it to the wrong side of the fabric. Then cut out the fabric, adding a $\frac{1}{4}$-inch seam allowance on all sides.

To construct the quilt, think of the large triangle as being composed of four smaller triangles. Each of the four triangles will be constructed separately. Create the center section first. Cut a large equilateral triangle with an altitude of 13 inches from the yellow fabric, adding a $\frac{1}{4}$-inch seam allowance. Then sew 1-inch-high triangles together in three strips of alternating black (14) and yellow (13) triangles. Each strip will start and

end with a black triangle. Sew one strip to each side of the large yellow triangle. Then sew a red parallelogram, made from two small equilateral triangles, to each of the three corners.

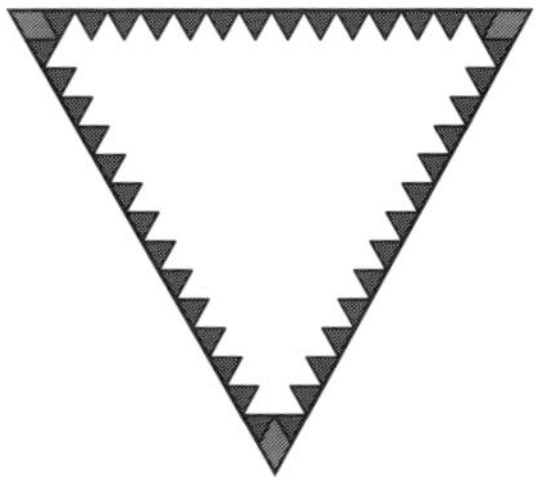

The three corner sections of the quilt are identical. To construct each section, break down each large triangle into smaller units and then piece them together. Use the drawing as a guide. When you have completed the three corner sections, attach them to the center section.

Add two triangles of black fabric to make a 50-by-42-inch rectangle. You can add a border made from one of the brighter fabrics. Layer the quilt top with batting and backing fabric. Baste the layers together with pins or large stitches. Use a quilting design that highlights the small equilateral triangles. In the quilt shown, black thread was used to quilt the large yellow triangles. You can use masking tape to form a grid on the fabric. Tape the masking tape onto the quilt with its edges following the sides of each of the individual triangles, forming a triangular grid. Stitch along the masking tape edges to emphasize all three sides of the triangles. Stitch in the "ditch," or seamline, of the pieced triangles and continue the triangle design over the remainder of the quilt. If you want to space the quilting lines farther apart, quilt every other row of triangles instead of every row.

Bind the outer edges of the quilt, and add a sleeve to the back for a rod to hang the quilt.

The Mascheroni Cardioid Quilt

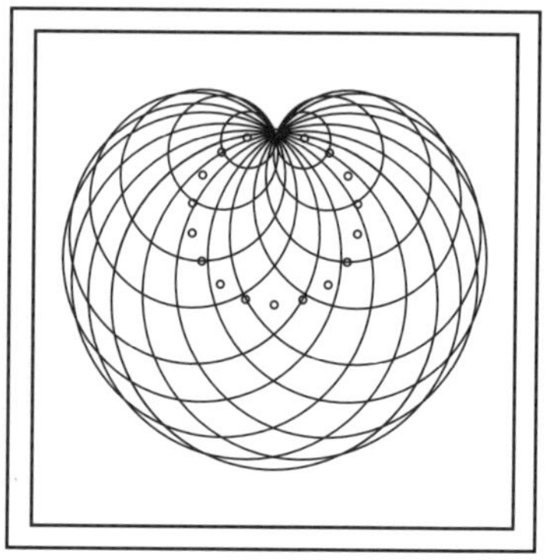

The Mascheroni cardioid is a heart-shaped figure created using only a compass. The figure is named for an eighteenth-century Italian mathematician, Lorenzo Mascheroni, known for compass-only constructions.

The Mascheroni cardioid is interesting because once an initial circle is drawn, the points on that circle used for the centers of the other circles are selected by the individual doing the construction. The shape of the cardioid outline is slightly different when it is created using different points on the circle.

The Mascheroni Cardioid quilt includes seventeen circles of different radii whose centers are evenly spaced around the circumference of the central circle. These circles outline a cardioid. The colors in the quilt graduate from light yellow to dark purple. The gradation of colors allows the viewer's eye to move around the cardioid with ease.

Designing Your Quilt

1. To create the Mascheroni cardioid, you will need a large sheet of paper and a compass. Follow the steps to construct the cardioid.

Step 1 Place the paper so that the long side is horizontal. Mark a point $2\frac{3}{4}$ inches from the top in the middle of the paper. Label the point *X*.

Step 2 Construct a circle with a radius of about $1\frac{1}{2}$ inches, centered at point *X*. (Open a compass to span $1\frac{1}{2}$ inches. Then, putting pressure on the point to keep it from moving and letting the pencil touch the paper, rotate the compass to draw a circle.)

Step 3 Mark point *A* at the top of the circle. Decide the number of circles you will use for the cardioid design. (Use an even number of intervals—at least 14. For the quilt design, 18 points were used.) Determine the angle that will mark circle *X* into equal intervals by dividing 360 degrees by the number of intervals.

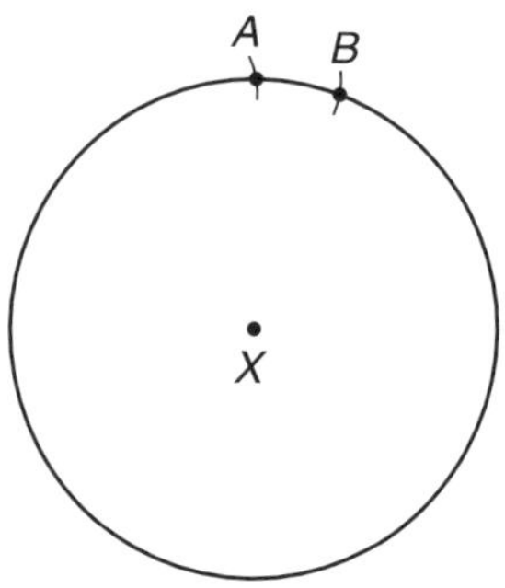

Use the quotient to determine the number of degrees of arc between equally spaced points. Mark point *B* that number of degrees away. (For 18 points, the arc is 20°.)

Step 4 Using *B* as the center and line segment *BA* as the radius, construct circle *B*. (That is, put the sharp end of your compass on point *B* and adjust the opening until the pencil end is on point *A*. Then turn the compass to create a circle without letting the opening on the compass change.)

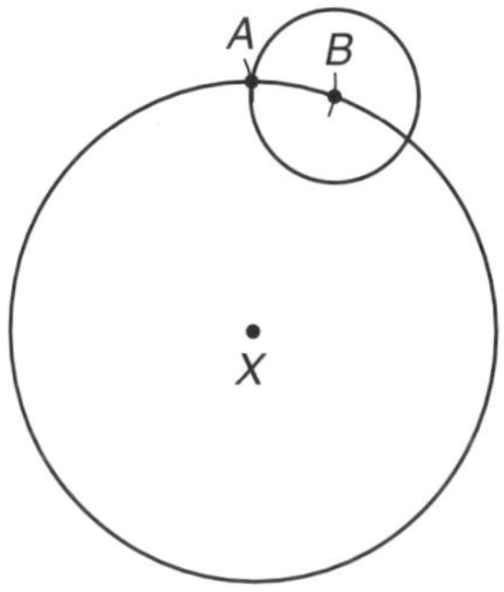

Step 5 Place another point *C* on circle *X* such that the distance from *B* to *C* is equal to the distance from *B* to *A*.

Step 6 Leaving the compass at the same setting (the distance from *B* to *A*), mark this interval all around the circle so you have between 14 and 18 points. Label them *D*, *E*, *F*, and so on.

Step 7 Using point *C* as the center and *CA* as the radius, construct circle *C*.

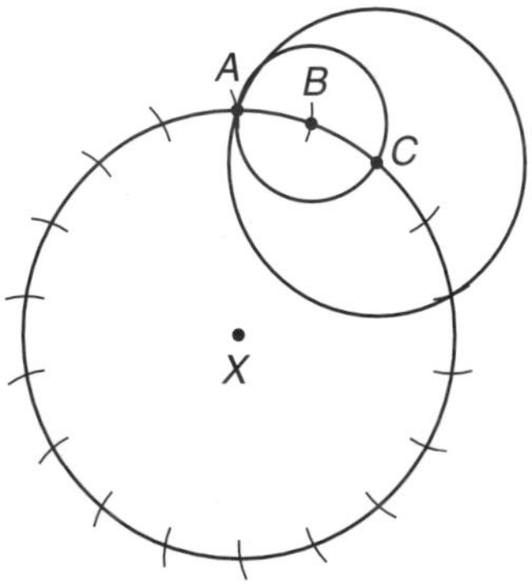

Step 8 Using points *D*, *E*, . . . , which are on the circle, draw circles centered at *D*, *E*, . . . with radii *DA*, *EA*, . . . , respectively. All the circles should go through *A*.

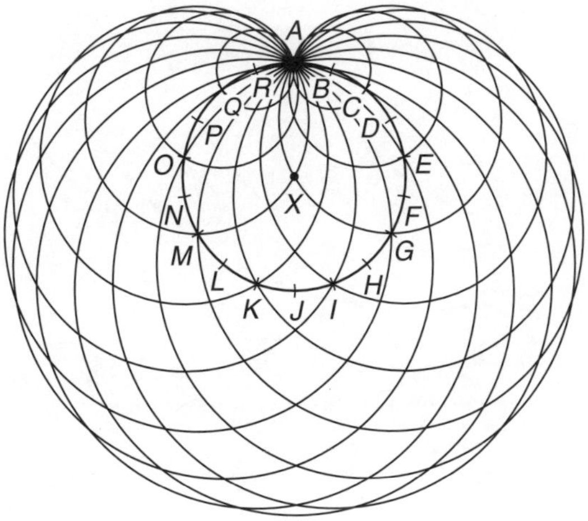

2. Color your design.

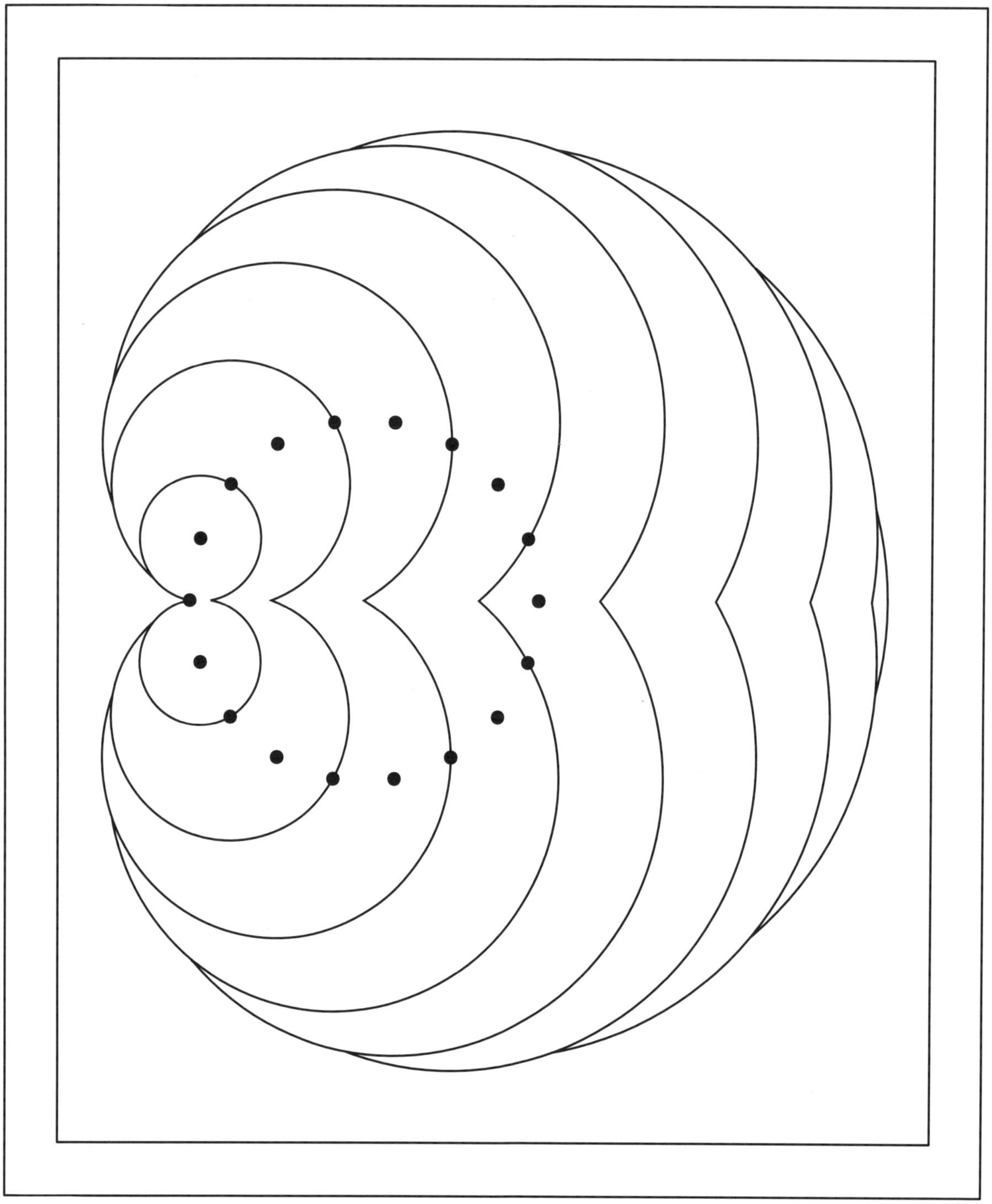

Making the Mascheroni Cardioid Quilt

The size of the finished quilt is 31 inches by 35 inches.

Materials

1 yard of light background and binding fabric

A variety of colorful fabrics to be used for the cardioid design

1 yard of backing fabric

17 beads with diameters of about $\frac{1}{4}$ inch

Freezer paper

Instructions

To create a pattern for the quilt, draw the cardioid full size on a large sheet of freezer paper. The original circle (circle *X*) should have a radius of exactly 5 inches. To determine the shape and placement of the 18 crescents and to ensure that the beads are at the centers of the 17 circles, follow these instructions to mark points *A* through *R*.

Step 1 At the top of circle *X*, draw a tangent line. Label the point where the line intersects the circle *A*.

Step 2 Mark off 10-degree angles with vertex *A* as shown. (These angles are equivalent to 20-degree central angles.)

Step 3 Mark and label the points where the 10-degree rays intersect the circle as shown.

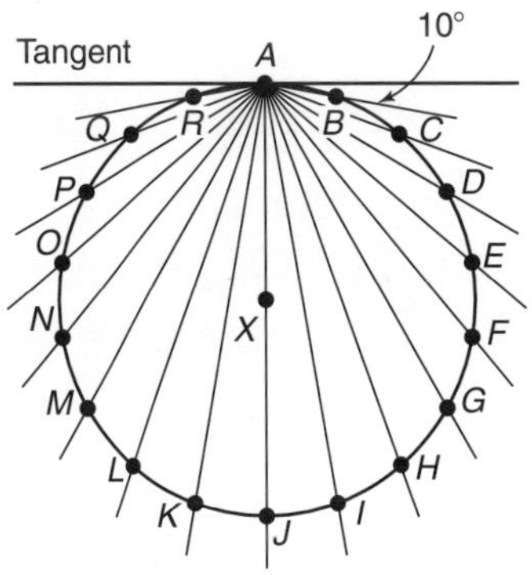

Step 4 For each point, starting with point *B*, draw a circle with center at the point and the distance from that point to point *A* as the radius (circle *B* with radius *BA*, circle *C* with radius *CA*, and so on).

To place the circle pieces correctly, transfer point A and the 17 circle centers to the background fabric. Cut a paper pattern circle for each of the 17 circles and mark the center, the horizontal diameter, and the placement of each fabric circle piece relative to the horizontal. Use these circles and the circle centers marked on the background fabric to correctly place each piece of colored fabric to the background.

Cut two mirror images of each of the eight parts of circles (with diameters 2 in., 3 in., 4.5 in., 6 in., 7 in., 8.25 in., 8.75 in., and 9 in.), each from a different color fabric. Cut one 9.5-in. circle from yet a different color fabric. Allow a little extra fabric on the side adjacent to other circles to allow for overlapping when the circles are appliquéd to the background. (Note: You can cut circles that will be trimmed in a later step or create the partial circle shapes that are visible in the quilt to save material.) Use the paper pattern as your guide to the shape of each piece.

Start at the bottom with the piece of the largest circle and appliqué it to the background. The second partial circles are appliquéd to the first circle and the foundation. If necessary, trim excess fabric before attaching the next partial circle. Continue until all the circles are appliquéd to the previous pieces and the background.

Trace the quilting lines around the circumference of each original circle using the circle patterns and a washable marking pencil or sewing chalk. Instead of drawing lines with a pencil, you can mark small pencil dots to outline the circles. Mark the lines for quilting with pins. A fairly heavy thread should be used to quilt along these circles.

After the quilting is finished, sew on the beads at the centers of the 17 circles. Add a binding around the border and a rod pocket for hanging the quilt.

Drawing Without a Compass

In designing the Mascheroni Cardioid quilt, you created a cardioid shape using a compass. You can also create a cardioid by rolling a circle around another circle.

1. To create the cardioid, you will need two circles with 1-inch radii, cut from tag board or stiff paper.

   Step 1 Secure the first circle to the center of a blank sheet of paper.

   Step 2 Mark a point on the edge of the other circle.

   Step 3 Roll the free circle completely around the secured circle, keeping the circles tangent. As you roll the circle, draw the path of the marked point as it moves.

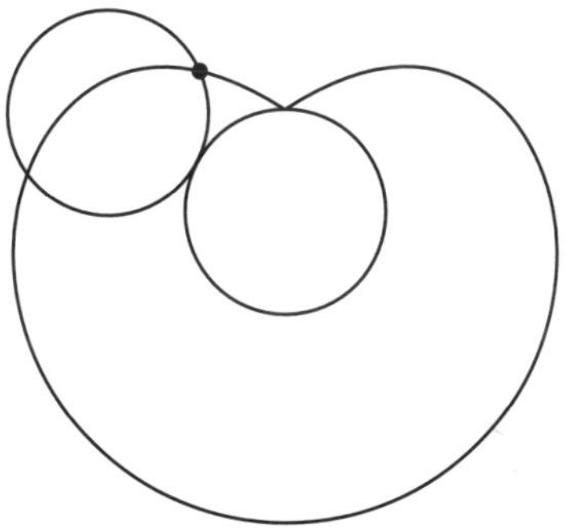

2. Compare the cardioid you drew in part 1 to the one you outlined using a compass. How are the designs the same? How are they different?

Mascheroni Construction 1

In a Mascheroni (compass-only) construction, the line segments are not drawn; they are imagined.

1. Draw and label two points, *A* and *B*. Imagine a segment whose endpoints are these two points. You can use a compass to construct a segment twice as long as segment *AB*.

   Step 1 Put the sharp end of your compass on point *A*. Adjust the opening so that the pencil end is on point *B* and swing a very full arc as shown here.

   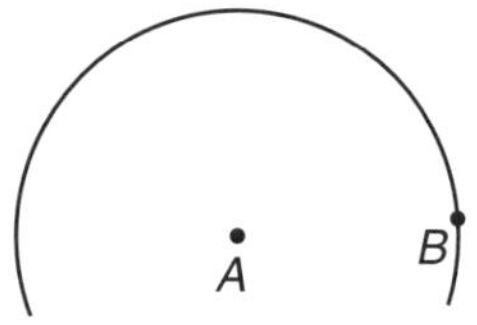

   Step 2 Without changing the opening of your compass, swing an arc from point *B* and label the point where it crosses the original arc *C*.

   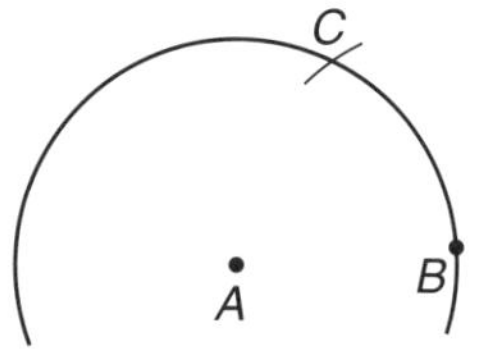

   Step 3 Without changing the opening of your compass, swing an arc from point *C* and label the point where it crosses the original arc *D*.

   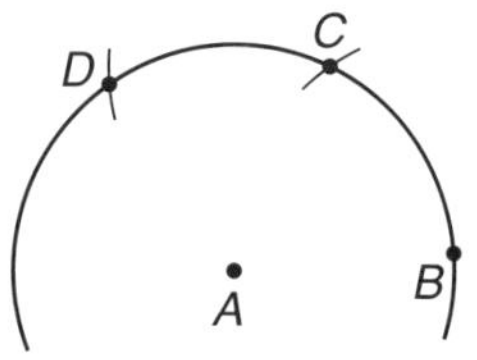

   Step 4 Without changing the opening of your compass, swing an arc from point *D* and label the point where it crosses the original arc *E*.

   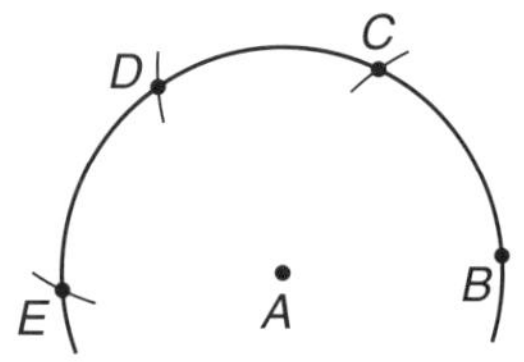

2. *BE* is a segment (*A*, *B*, and *E* are collinear) twice as long as segment *AB*. That is, $BE = 2AB$. Explain why this is true, using the fact that triangles *AED*, *ADC*, and *ABC* are equilateral.
3. Explain and show how you could continue the construction above to create a segment three times as long as segment *AB*.
4. Show a very simple way to draw a segment twice as long as *AB* using a compass and a straightedge.

ACTIVITY 3

Mascheroni Construction 2

In a Mascheroni (compass-only) construction, you must imagine the segments between points.

1. Construct a segment that is $\frac{1}{3}$ the length of a given segment.

Step 1 Draw and label two points, *A* and *B*. You will construct a segment that is $\frac{1}{3}$ the length of segment *AB*.

Step 2 Use Mascheroni Construction 1 (see Activity 2) to find a point *C* so that $AC = 3AB$.

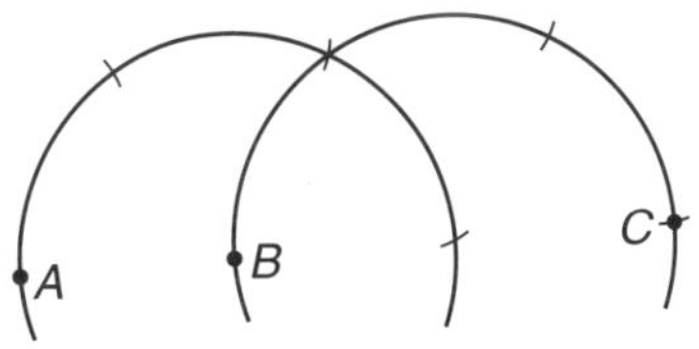

Step 3 Draw a circle with center *A* and radius *AB*.

Step 4 Put the sharp end of your compass on point *C* and construct an arc with radius *AC*. Make the arc large enough so that it intersects circle *A* in two points. Label the points of intersection *D* and *E*.

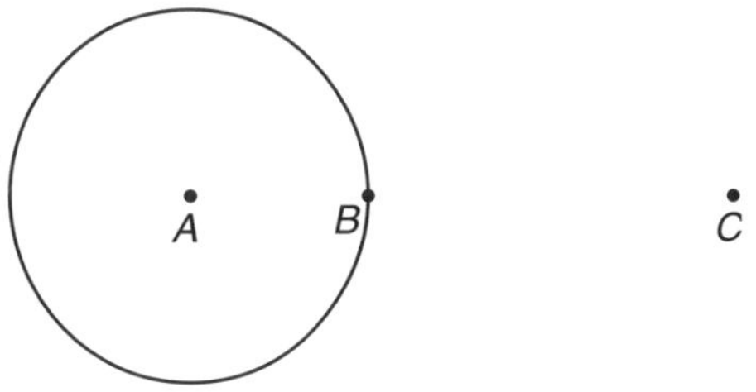

Step 5 Put the sharp end of your compass on point *D* and construct an arc with radius *DA*. Then put the sharp end of your compass on point *E* and construct an arc with radius *EA*. Label the point where the arcs cross *F*.

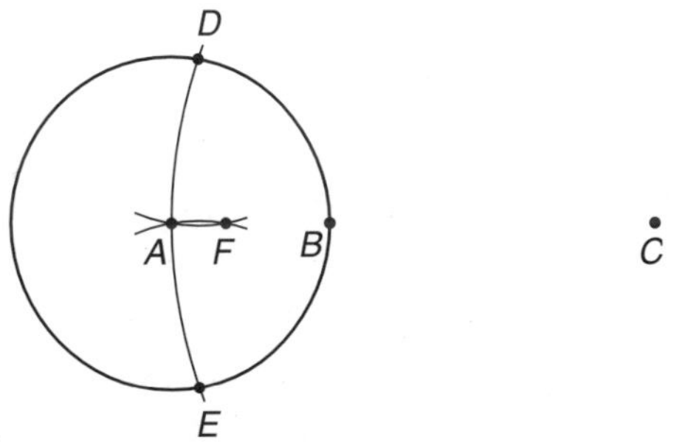

The length of segment *AF* is $\frac{1}{3}$ the length of *AB*. It is possible to justify this conclusion by looking at similar triangles *DAF* and *CDA* and the ratio of their corresponding sides.

2. Explain how you know the corresponding angles of these triangles are equal and therefore know the triangles are similar.

3. How does knowing they are similar triangles help you prove that the length of segment *AF* is $\frac{1}{3}$ the length of *AB*?

Constructing a Regular Hexagon

1. Use a compass and straightedge to construct a regular hexagon.

Step 1 Draw a circle with a radius of about 3 inches, and mark a point *A* anywhere on the circle.

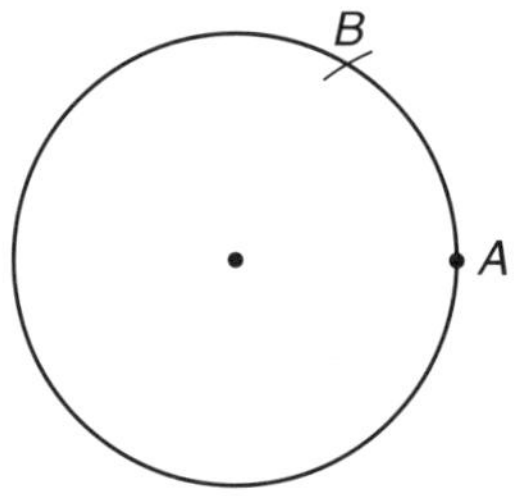

Step 2 Without changing the opening, place the sharp end of your compass on point *A* and draw an arc across the circle. Label the point where the arc crosses the circle *B*.

Step 3 Without changing the opening, place the sharp end of your compass on point *B* and draw an arc on the circle. Label the intersection point *C*.

Step 4 Continue this process of placing the sharp end of the compass on the last point marked, striking an arc, and labeling the point of intersection. The sixth arc will bring you back to point *A*.

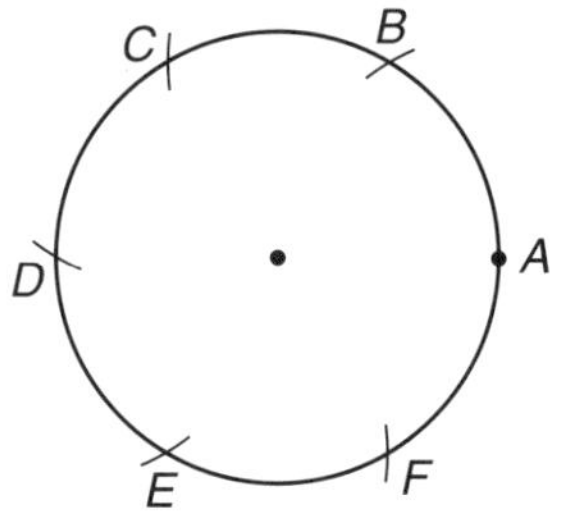

Step 5 Draw segments connecting points *A* to *B*, *B* to *C*, *C* to *D*, *D* to *E*, *E* to *F*, and *F* to *A* to form a regular hexagon.

2. To create a regular dodecagon (12-sided polygon), mark six equal arcs on a circle and find the midpoint of one of the segments that is one side of a hexagon. Draw a diameter of the circle through that midpoint, and use the intersection of the diameter with the circle as a starting place. Using the compass opening equal to the radius of the circle, mark five more points. Connect the 12 points in order.

3. Early compass constructions were done by drawing full circles instead of arcs. Follow the steps for creating a regular hexagon, but make full circles instead of arcs. Describe the design you see. How many colors would you need to color the design so that any two sections sharing an edge are different colors?

ACTIVITY 5

The Clamshell and the Ax

This shape is called a *clamshell.*

And this is called an *ax.*

1. Trace one of these shapes in the center of a sheet of tracing paper. Then move the paper so the tracing fits together with the original shape and trace the original shape again.

   Continue tracing the shape to create a *tessellation*—copies of a shape that cover a surface without gaps or overlaps.

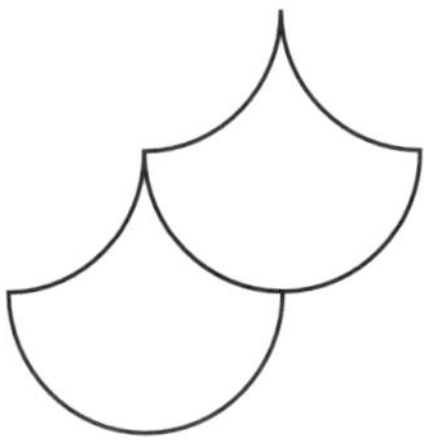

2. See if you can find a way to cut the clamshell or the ax on Worksheet 18 into pieces that can be rearranged to form a rectangle. Describe the cuts you made and how you fit the pieces together.
3. This shape, called a hinged dissection, is related to the clamshell. Use Worksheet 18 to experiment with cutting it apart and reassembling the pieces into a rectangle.

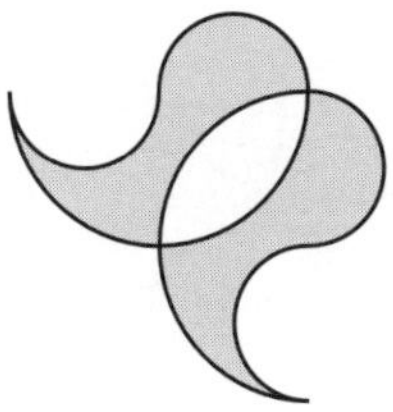

WORKSHEET 18

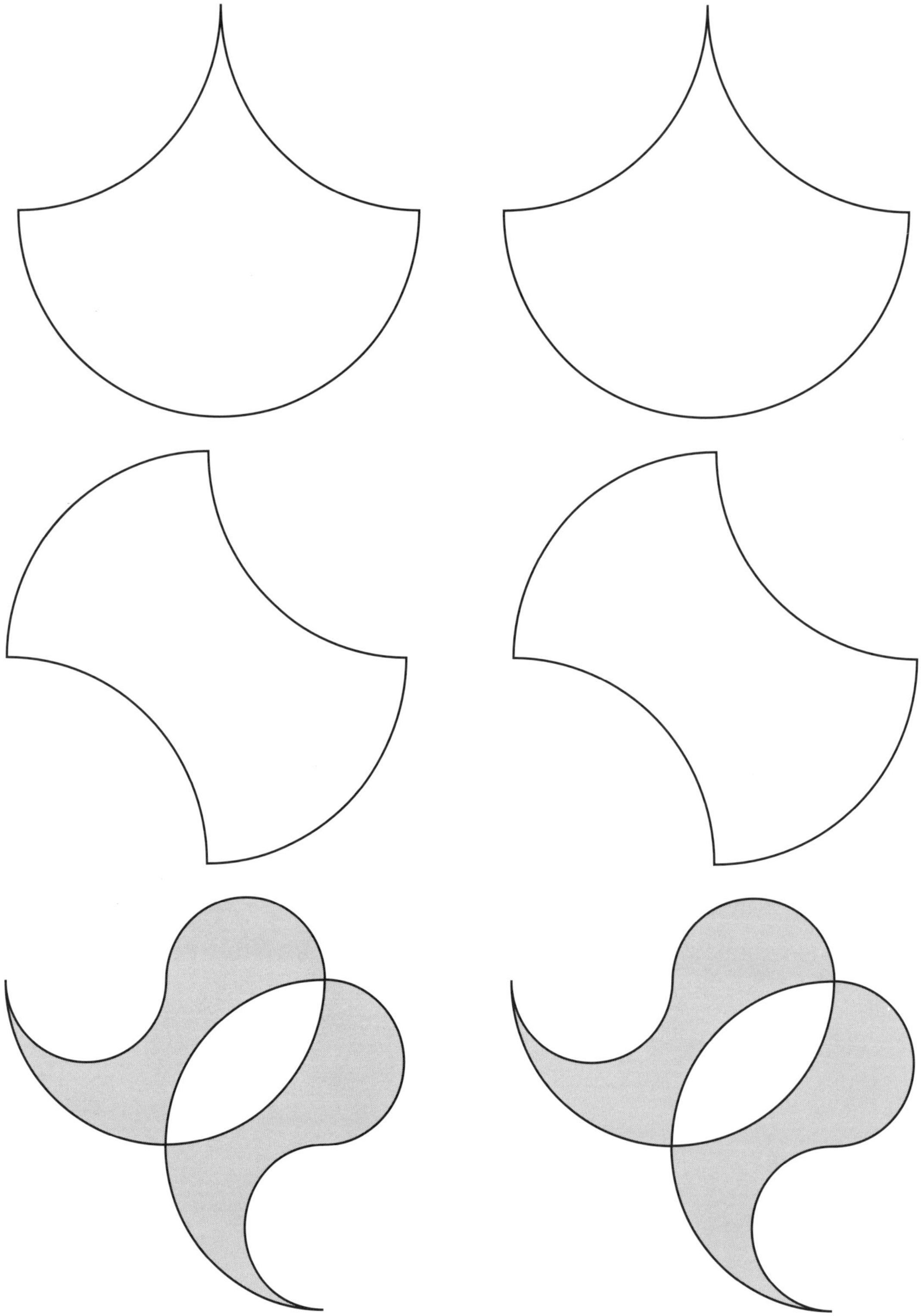

ACTIVITY 6

Leonardo's Ellipse

Leonardo da Vinci discovered an interesting way to draw an ellipse.

1. To draw an ellipse, you will need a plain sheet of paper, a straightedge, a 3-by-5-inch card, and a pencil.
   a. On a plain sheet of paper, draw two perpendicular axes that intersect near the center of the paper.
   b. On a 3-by-5-inch card, draw any triangle and cut it out. Label the vertices *A*, *B*, and *C*.
   c. Place the triangle on the sheet of paper so that vertex *A* touches one axis and vertex *B* touches the other. Mark a point showing the location of vertex *C*.

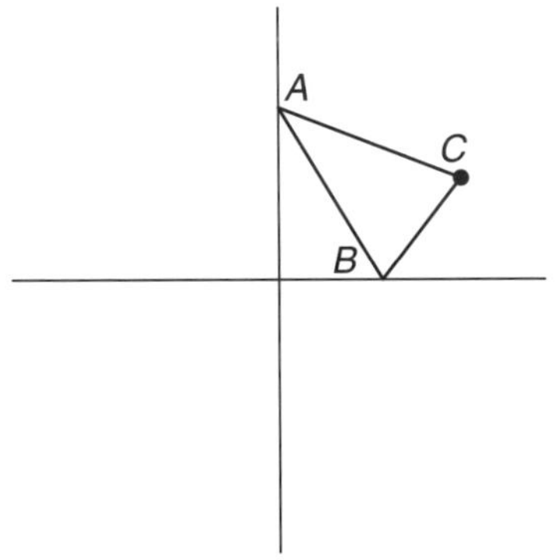

   d. Move the triangle slightly, sliding vertex *A* along one axis and vertex *B* along the other. Mark the new location of point *C*.
   e. Continue moving the triangle in this way—making sure vertices *A* and *B* always touch the axes—and marking the location of vertex *C*. The path (or locus of all the points) of vertex *C* will be an ellipse.
2. Repeat the process in part 1 with axes that are not perpendicular. Do you still get an ellipse?

Sums and Patterns

On the copy of Pascal's triangle from Worksheet 19, number the rows from 0 to 16. Add the numbers in each row and record the sum to the right of the row.

row 0	1	1
row 1	1 1	2
row 2	1 2 1	___
row 3	1 3 3 1	___
row 4	1 4 6 4 1	

1. What is the sum of the numbers in row 17?
2. Describe the pattern in the sums.
3. How could you find the sum of the numbers in row *n* without adding them?
4. What is the sum of the numbers in row 50?

Use several copies of Pascal's triangle to shade patterns.

5. Shade the even numbers on Pascal's triangle.
6. Shade multiples of 3.
7. Shade multiples of 4.
8. Shade multiples of 5.
9. Describe the patterns you see in each shading.

WORKSHEET 19

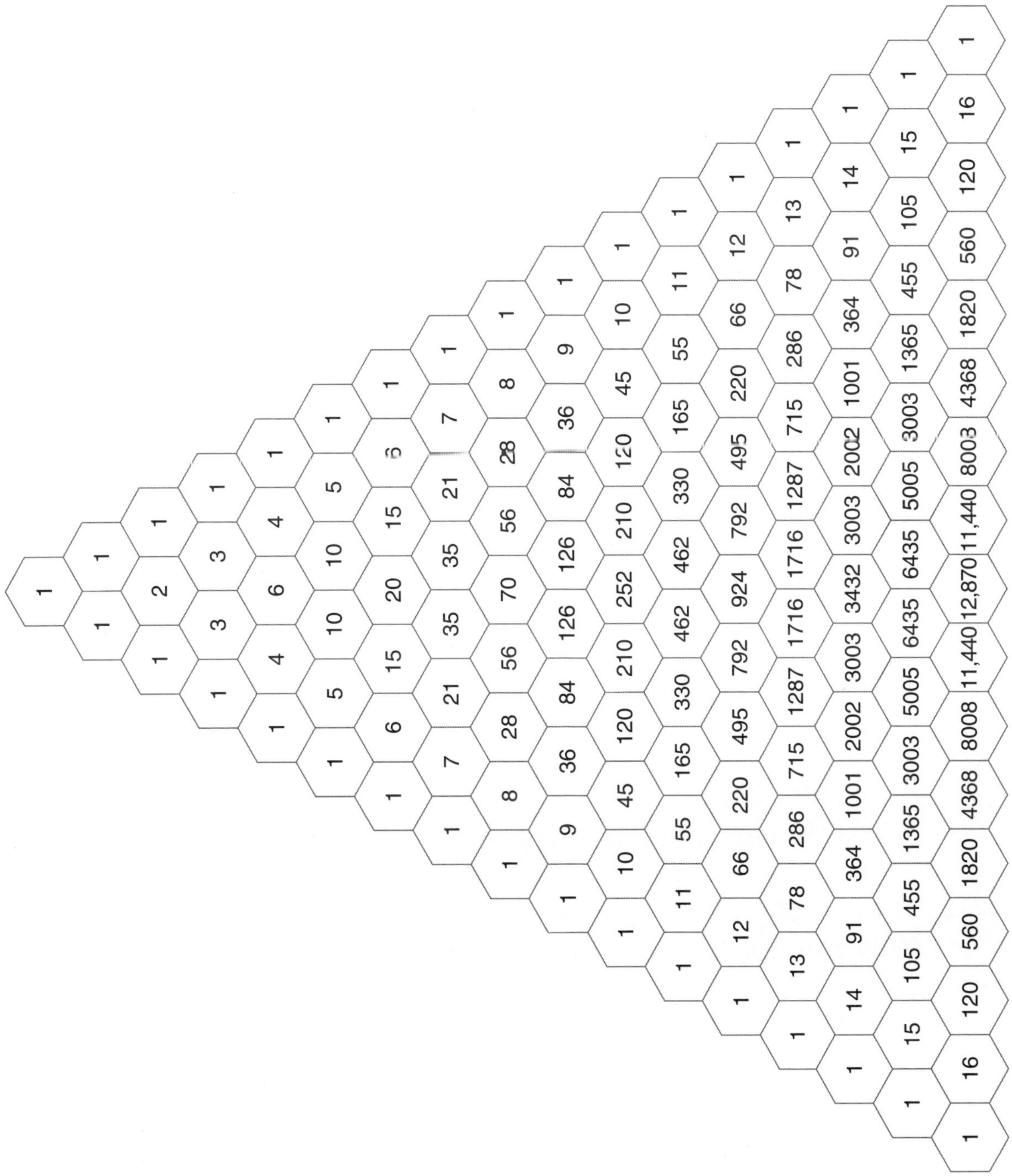

Finding Figurate Numbers in Pascal's Triangle

Figurate numbers can be represented by polygons of dots.

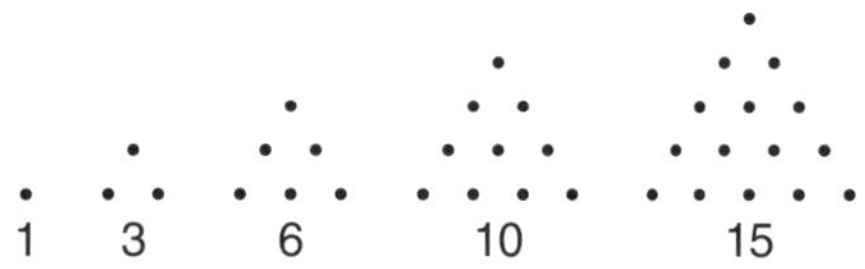

Triangular numbers

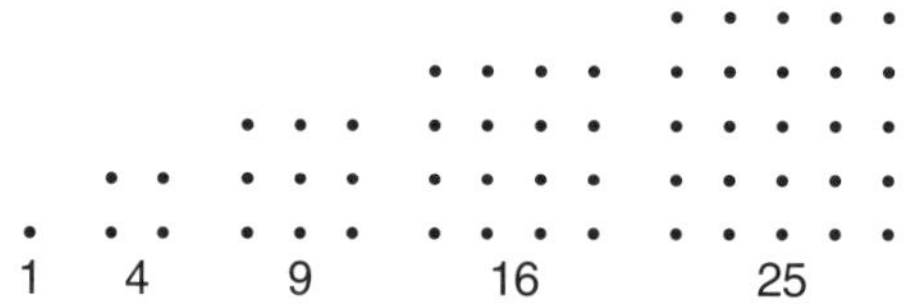

Square numbers

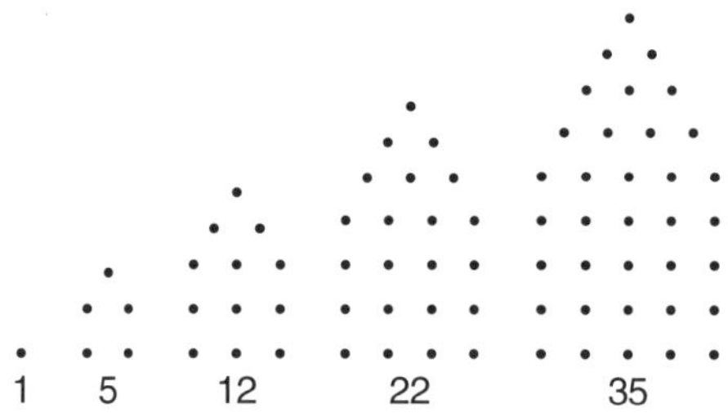

Pentagonal numbers

For this activity, you will need three copies of Pascal's triangle from Worksheet 19.

1. List the next four triangular numbers. On one copy of Pascal's triangle, shade the triangular numbers.
2. List the next four square numbers. On another copy of the triangle, shade the numbers whose sums are square numbers.
3. List the next four pentagonal numbers. Shade the numbers that sum to the pentagonal numbers. Use the triangular numbers as a basis. What patterns do you see?

Pascal's Triangle and Coin Tossing

Binomial probabilities in which there are several trials of an event with two possible outcomes for each trial can be determined from Pascal's triangle. The probabilities for possible outcomes when tossing three coins can be seen in the third row of Pascal's triangle. The numbers are 1 3 3 1. There are eight possible outcomes: HHH, HHT, HTH, HTT, THH, THT, TTH, and TTT. Of the eight choices, one represents getting three heads. Three choices involve two heads and a tail. The table shows the probability of each number of heads for the eight possible outcomes.

Tossing Three Coins

Number of heads	3	2	1	0
Number of ways	1	3	3	1
Probability	$\frac{1}{8}$	$\frac{3}{8}$	$\frac{3}{8}$	$\frac{1}{8}$

All of the eight outcomes are equally likely. This means that the probability of getting two heads and one tail is $\frac{3}{8}$, or 37.5 percent. The probability of two tails and one head is the same.

1. Using the fourth row of Pascal's triangle, 1 4 6 4 1, shows the possible outcomes from tossing four coins.

Tossing Four Coins

Number of heads	4	3	2	1	0
Number of ways					
Probability					

2. Explain why the probability of getting one tail with three heads is the same as three tails with one head.
3. What is the probability that the four tossed coins will all come up heads?

Research Activities

1. During the Renaissance, many mathematicians attempted to square the circle. Research some of the interesting attempts to square the circle.
2. Michael Stifel (1486–1567) was one of the most important German algebraists of the sixteenth century. Research Stifel's work on Pascal's triangle. How does his work relate to the work that the Chinese did centuries earlier?
3. Research and report on any of the following topics.

   cardioid

   compass constructions

   Dürer, Albrecht

   fixed-compass constructions

   Mascheroni, Lorenzo

   Mohr, George

   Pascal, Blaise

   Pascal's triangle

   Plato

   Poncelet, Jean Victor

   Steiner, Jakob

Activities 1–4 extend the ideas of the Mascheroni Cardioid quilt. You may wish to share some of the history of compass constructions discussed on page 131 to explain why this quilt pattern was named for the mathematician Mascheroni. In Activity 1, students learn a different way of creating a cardioid shape—by rolling a circle around another congruent circle. Instead of tag board circles, you can use quarters. Or use a graphing calculator to generate a cardioid. Check your calculator's manual for the instructions.

In the next two activities, students learn basic compass-only constructions. The experience of creating constructions enhances students' visual and spatial skills; it helps them understand illustrations and see relationships that they can use to justify conjectures. If your students have little experience with compasses, discuss and demonstrate compass-and-straightedge constructions before having students do these activities. Emphasize the importance of holding the compass by the small handle at the top rather than by its sides. Students apply construction skills to create polygons and a design in Activity 4.

The design used in the Leonardo's Dessert quilt is based on Leonardo's work with curved shapes. In Activity 5, discuss the idea of finding a figure with straight sides that has the same area as a curved figure. In Activity 6, students learn a method Leonardo discovered for generating an ellipse.

Before students begin their designs for the Pascal's Surprise quilt, review the material on mod-10 arithmetic on page 138. Have triangular grid paper on hand to help students, and encourage them as they explore patterns and create their own designs using the triangle. Mathematicians look for and study patterns; there are always new patterns to be discovered. Marc Roth started with Pascal's triangle, wrote its units digits in a grid of triangles, added numbers around the grid's empty triangles, and added color to reveal a pattern in the digits.

Activities 7–9 extend the ideas of Pascal's triangle. First, students explore the pattern in the sums of the rows and look for visual patterns as they shade multiples of 2, 3, 4, and 5 in the triangle. In Activity 8, students find patterns as they look for numbers that sum to the figurate numbers. Activity 9 shows a relationship between outcomes of coin tosses and the numbers in a row of Pascal's triangle.

Some activities and designs can be done in groups. For example, each of four students can create two hinged dissections, and the group can work together on the final design.

Many interesting topics are related to the quilts in this section. Activity 10 suggests ideas for research.

TEACHER NOTES

Answers and Comments

Activity 1

If your students are familiar with polar equations, you can have them investigate the polar equation of a cardioid: $r = 2a(1 + \cos\theta)$ or $r = 2a(1 - \cos\theta)$, where a is any real number. The length of the cardioid is $2a$, and the area enclosed by the cardioid is $6\pi a^2$. The length of the cardioid is also the diameter of the largest circle drawn in constructing the Mascheroni cardioid.

1. 

2. The cardioid constructed with the compass shows individual circles whose outside edges outline a cardioid figure. The same figure made in this activity shows the entire outer edge of the cardioid.

Activity 2

Deriving the proof in this construction exercise is appropriate for geometry students, but younger students might be able to understand the proof if you talk it through with them.

2. Because $\triangle AED$, $\triangle ADC$, and $\triangle ABC$ are equilateral triangles, each angle has the same measure:

$$m\angle EAD = m\angle DAC = m\angle CAB = 60°$$

Therefore, the $m\angle EAD + m\angle DAC + m\angle CAB = 180°$, so points A, B, and E are collinear.

Because they are both radii of the circle, $AE = AB$; therefore, $BE = 2AB$.

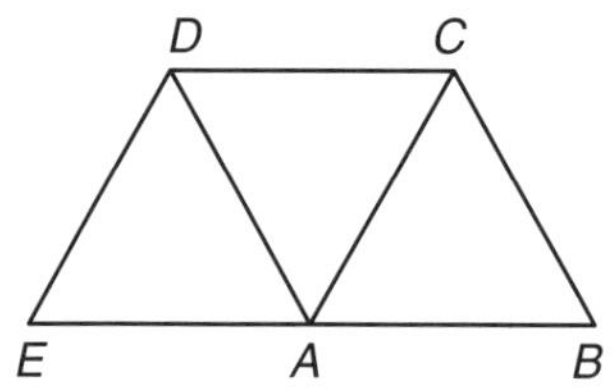

3. Start by constructing segment *AE*. Then repeat the construction to find a point *F* so that $\overline{AF}$ is twice $\overline{AE}$. The length of $\overline{BF}$ will be three times the length of $\overline{AB}$.

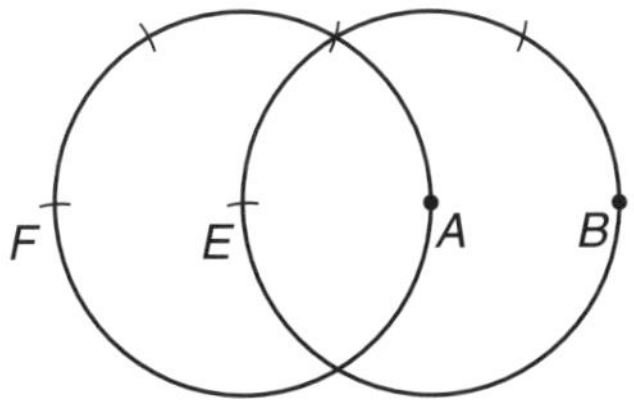

4. Extend the line segment beyond *B*, and then, putting the compass point on *B*, mark on that segment an arc of radius *AB*.

Activity 3

2. Because $\overline{CD}$ and $\overline{CA}$ are radii of the same circle, they are congruent. Therefore, $\triangle ADC$ is an isosceles triangle with $\angle A \cong \angle D$. Similarly, $\triangle FAD$ is isosceles, and $\angle A \cong \angle AFD$. Because $\angle A \cong \angle A$, $\angle D \cong \angle AFD$, $\triangle ADC$ is similar to $\triangle FAD$.

3. Because $AB = AD$ and corresponding parts of similar triangles are proportional, $AB = \frac{1}{3}AC$, so $AD = \frac{1}{3}AC$. Because the sides of $\triangle FAD$ are in the same proportion as the corresponding sides of $\triangle ADC$, $AF = \frac{1}{3}AD$. Because $AD = AB$, $AF = \frac{1}{3}AB$.

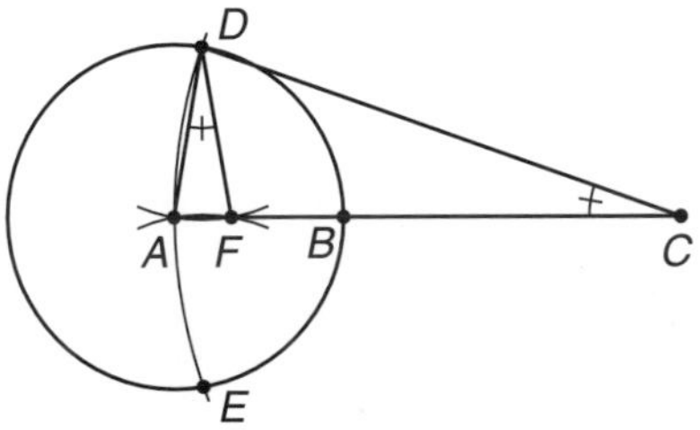

Activity 4

1.

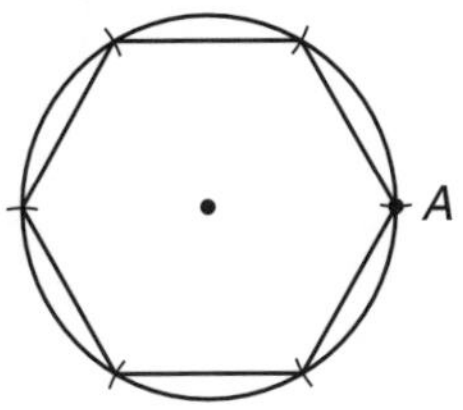

2.

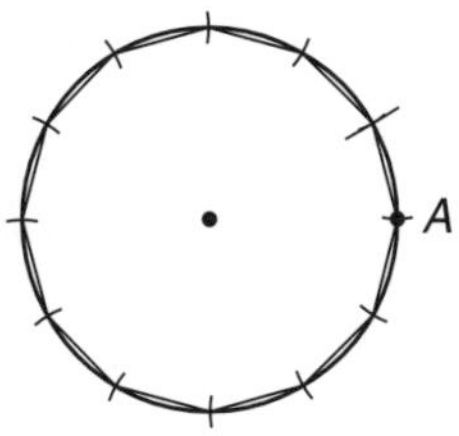

3.

Students might describe the design as similar to a flower, as symmetrical (with both rotational symmetry and six lines of reflection), or as showing six circles sharing one vertex. The design can be colored with just two colors.

As an extension, you could introduce the map-color problem: How many colors are needed to color any map drawn on a flat surface? It was proved in the mid-1900s that at most four colors are needed to color any map. That is, it is impossible to draw a map that would require five colors. Of course, five or more colors could be used to color a map, but only four are needed.

Activity 5

1. Possible answer:

2. Clamshell: Cut straight across at the widest point, and cut the top piece in half through its line of symmetry. Then fit the two top pieces along the curves on the bottom piece.

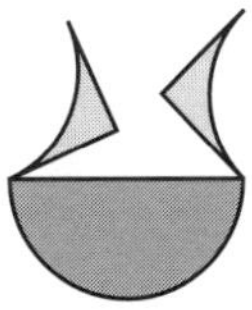

Ax: Cut into four pieces and rearrange as shown.

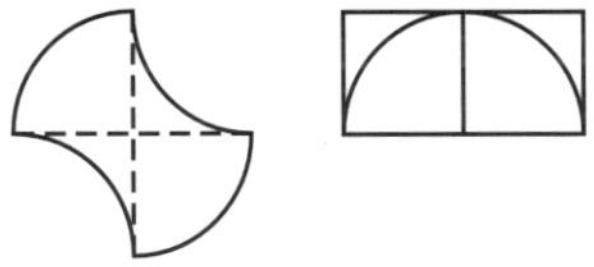

3. Hinged dissection: Make four cuts as shown, and fit the six pieces into a rectangle.

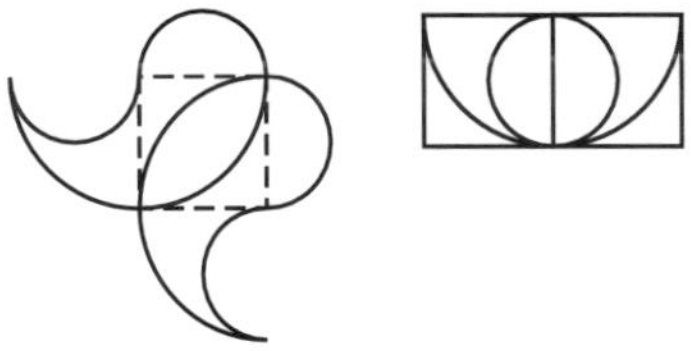

Activity 6

1. Sample answer for this triangle:

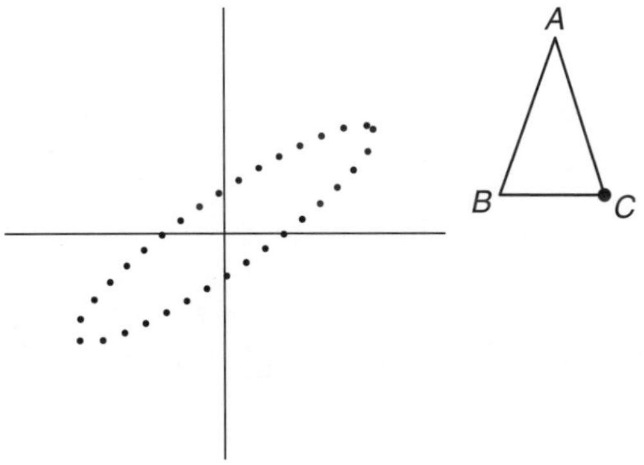

2. Yes, the path of the third vertex still traces out an ellipse.

TEACHER NOTES

Activity 7

Row	Sum
0	2^0 or 1
1	2^1 or 2
2	2^2 or 4
3	2^3 or 8
4	2^4 or 16
5	2^5 or 32
6	2^6 or 64
7	2^7 or 128
8	2^8 or 256
9	2^9 or 512
10	2^{10} or 1,024
11	2^{11} or 2,048
12	2^{12} or 4,096
13	2^{13} or 8,192
14	2^{14} or 16,384
15	2^{15} or 32,768
16	2^{16} or 65,536

1. 2^{17} or 131,072
2. Possible answer: The sum of each row is twice the sum of the row before it.
3. The sum of row n is 2^n.
4. 2^{50}

5. Multiples of 2

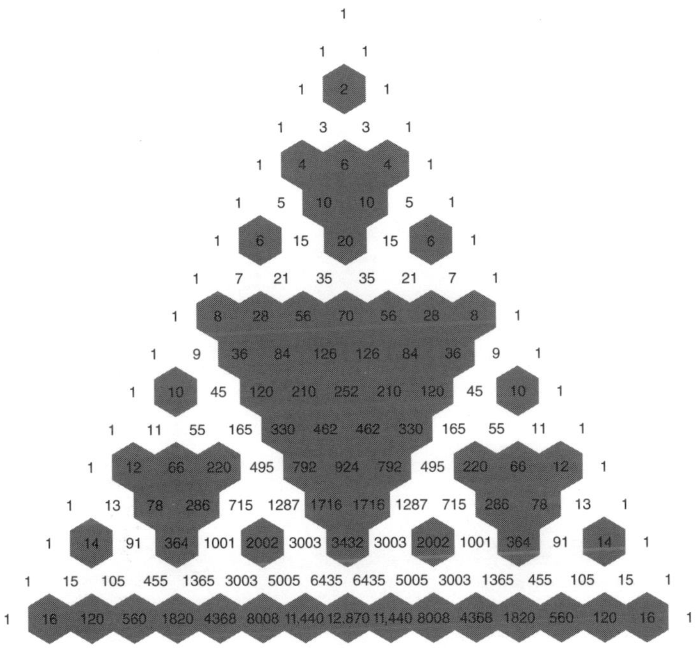

6. Multiples of 3

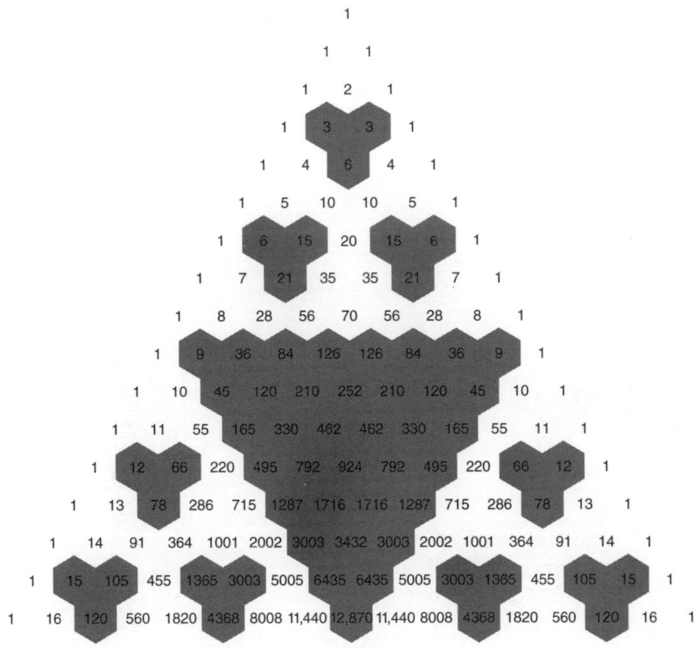

7. Multiples of 4

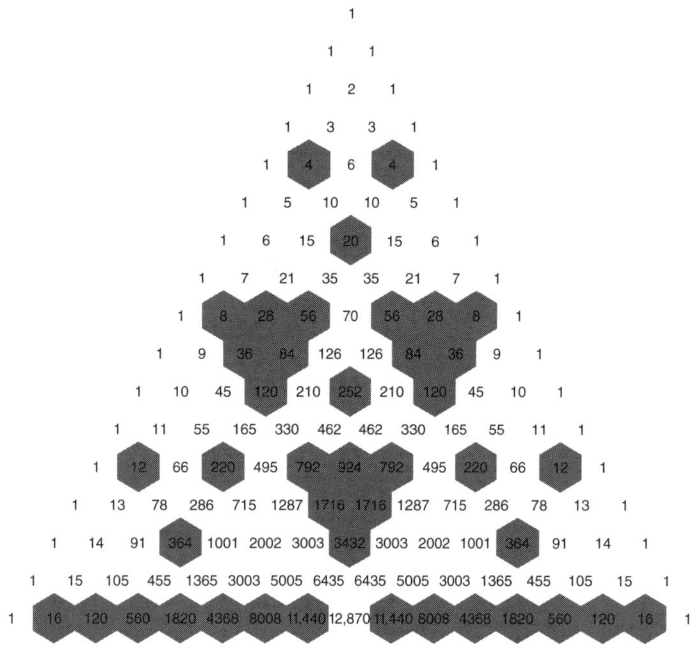

8. Multiples of 5

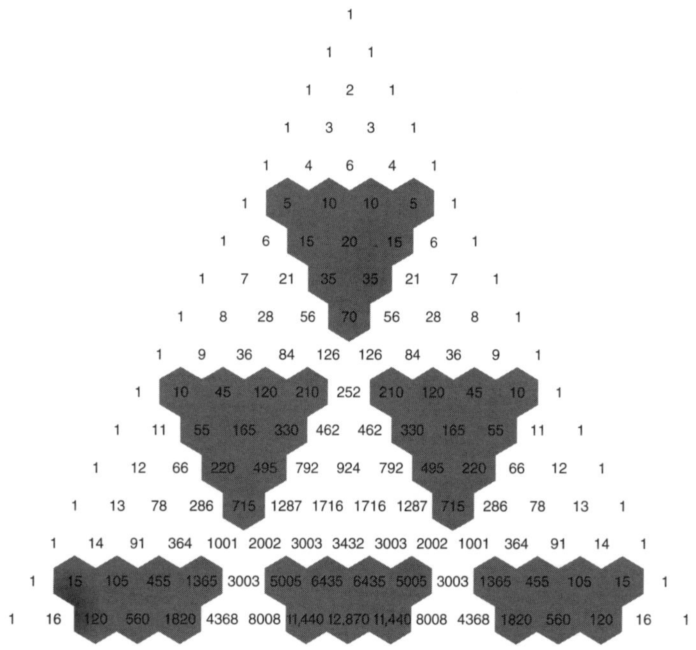

9. Answers might include:

   Each of the patterns has triangular-shaded regions.

   The shaded triangles all point downward.

   Small regions have the same patterns with smaller triangles as larger regions have.

   The multiples of 2 look like the pattern in Sierpiński's triangle with the smallest triangle having just one number shaded.

   The smallest shaded triangle in the multiples of 3 covers three numbers.

   The smallest shaded triangle in the multiples of 4 covers one number, and the next smallest covers six numbers; there are no patterns of just three shaded numbers together.

   The part of the multiples of 5 shaded here shows shaded triangles of 15 cells, and the next largest is

$$24 + 23 + 22 + 21 + \cdots + 4 + 3 + 2 + 1 = 1587.$$

Activity 8

1. 21, 28, 36, 45. The triangular numbers are found along the diagonal that contains the 1 in row 2, the 3 in row 3, the 6 in row 4, the 10 in row 5, ...

1
3
6
10
15
21
28
36
45
55
66

TEACHER NOTES

2. 36, 64, 81, 100. The square numbers can be found by adding two adjacent numbers in the diagonal that contains the triangular numbers to two numbers next to them.

	1
1	2
3	3
6	4
10	5
15	6
21	7
28	8
36	9
45	10
55	11
16	12

$1 + 2 + 3 + 3 = 9$

$3 + 3 + 6 + 4 = 16$

$6 + 4 + 10 + 5 = 25$

$10 + 5 + 15 + 6 = 36$

The square numbers can also be found by taking groups of four numbers from the diagonal that starts at row 2 and the diagonal that starts at row 2: $1 + 2 + 3 + 3 = 9$; $3 + 3 + 6 + 4 = 16$; $6 + 4 + 10 + 5 = 25$.

1
3
6
10
15
21
28
36
45
55
66

$1 + 3 = 4$

$3 + 6 = 9$

$6 + 10 = 16$

$10 + 15 = 25$

$15 + 21 = 36$

3. 51, 70, 92, 117. Start with the same diagonal (1, 3, 6, 10, . . .). Take the two numbers above and the one number below to the left and add them to the number in that diagonal to get the pentagonal number: $2 + 1 + 6 + 3 = 12$; $3 + 3 + 10 + 6 = 22$; $4 + 6 + 15 + 10 = 35$.

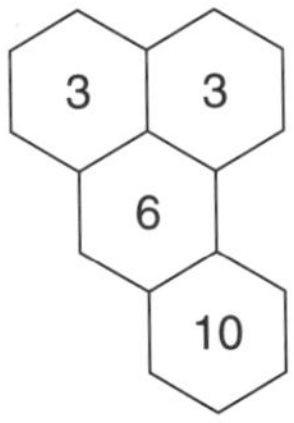

Sum is 22

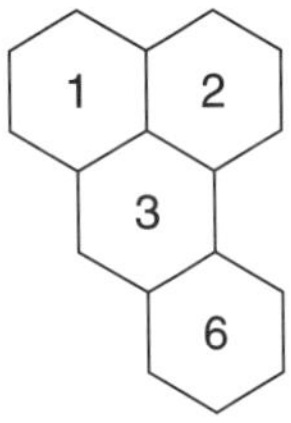

Sum is 12

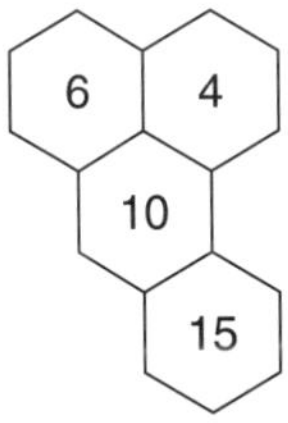

Sum is 35

1

1 1

1 2 1

1 3 3 1

1 4 6 4 1

1 5 10 10 5 1

6 15 20 15 6 1

21 35 35 21 7 1

56 70 56 28 8 1

126 126 84 36 9 1

252 210 120 45 10 1

462 330 165 55 11 1

792 495 220 66 12 1

1287 715 286 78 13 1

2002 1001 364 91 14 1

3003 1365 455 105 15 1

4368 1820 560 120 16 1

Teacher Notes

Activity 9

1. **Tossing Four Coins**

Number of heads	4	3	2	1	0
Number of ways	1	4	6	4	1
Probability	$\frac{1}{16}$	$\frac{1}{4}$	$\frac{3}{8}$	$\frac{1}{4}$	$\frac{1}{16}$

2. The probability of getting one tail with three heads is the same as three tails with one head, that is, $\frac{1}{4}$.

   Looking at the possible outcomes:
   HHHH; HHHT; HHTH; HHTT; HTHH*; HTHT; HTTH; HTTT**; THHH*; THHT; THTH; THTT**; TTHH; TTHT**; TTTH**; TTTT; *

   *Three heads with one tail
   **Three tails with one head

3. The probability of getting all heads when four coins are tossed is $\frac{1}{16}$.

Resources

Abbott, Edwin. *Flatland: A Romance of Many Dimensions.* New York: Dover Publications, 1952.

Ball, W. W. Rouse. *A Short Account of the History of Mathematics.* New York: Dover Publications, 1960.

Banchoff, Thomas F. *Beyond the Third Dimension—Geometry, Computer Graphics, and Higher Dimensions.* New York: W. H. Freeman and Co., 1990.

Barnsley, Michael F. *Fractals Everywhere.* Boston: Academic Press, Inc., 1988.

Choate, Jonathan, Robert L. Devaney, and Alice Foster. *Fractals: A Tool Kit of Dynamic Activities.* Emeryville, Calif.: Key Curriculum Press, 1999.

Cook, Theodore A. *The Curves of Life.* New York: Dover, 1979.

Dixon, Robert. *Mathographics.* New York: Dover, 1987.

Emmer, Michele, ed. *The Visual Mind: Art and Mathematics.* Cambridge, Mass.: The MIT Press, 1993.

Feder, Jens. *Fractals.* New York: Plenum Press, 1988.

Frederickson, Greg N. *Dissections—Plane and Fancy.* Cambridge, U.K.: Cambridge University Press, 1997.

Gleick, James. *Chaos: Making a New Science.* New York: Penguin, 1987.

Grattan, I., ed. *Companion Encyclopedia of the History and Philosophy of Mathematical Sciences, Volume 1.* London, U.K.: Routledge, 1994.

Green, Thomas, and Charles L. Hamberg. *Pascal's Triangle.* Palo Alto, Calif.: Dale Seymour Publications, 1986.

Hallerberg, Arthur E., ed. *Historical Topics for the Mathematics Classroom.* Reston, Va.: National Council of Teachers of Mathematics, 1989.

Hart, George W., and Henri Picciotto. *Zome Geometry.* Emeryville, Calif.: Key Curriculum Press, 2001.

Hess, Adrien L. *Four-Dimensional Geometry.* Reston, Va.: NCTM Publications, 1977.

Hockney, David. *Secret Knowledge: Rediscovering the Lost Techniques of the Masters.* New York: Penguin, Putnam, Inc., 2001.

Jacobs, Harold E. *Mathematics—A Human Endeavor,* 3d edition. New York: W. H. Freeman & Co., 1994.

Jones, Judy, and William Wilson. *An Incomplete Education.* New York: Ballantine Books, 1995.

Appendix

Kappraff, Jay. *Connections: The Geometric Bridge between Art and Science.* New York: McGraw-Hill, Inc., 1991.

Kline, Morris. *Mathematical Thought From Ancient to Modern Times.* New York: Oxford University Press, 1973.

Labella, Vincenzo. *A Season of Giants: Michelangelo, Leonardo, Raphael, 1492–1508.* Boston: Little, Brown & Co., 1990.

Lauwerier, Hans. *Fractals: Endlessly Repeated Geometrical Figures.* Princeton, N.J.: Princeton University Press, 1991.

Lénárt, István. *Non-Euclidean Adventures on the Lénárt Sphere.* Berkeley, Calif.: Key Curriculum Press, 1996.

Mack, Charlotte Williams. "Exploring Three- and Four-Dimensional Space." *The Mathematics Teacher.* October, 1995. pp. 572–575.

Manning, Henry P. *Fourth Dimension Simply Explained.* Magnolia, Mass.: Peter Smith Publications, 1950.

Morin, David R. *Perspective Made Easy.* Portland, Maine: J. Weston Walsh, 1985.

Mosovich, Ivan. *The Magic Cylinder Book.* Norfolk, England: Taravin Publications, 1997.

Pappas, Theoni. *The Magic of Mathematics.* San Carlos, Calif.: Wide World Publishing/Tetra, 1994.

Pappas, Theoni. *Mathematics Appreciation.* San Carlos, Calif.: Wide World Publishing/Tetra, 1986.

Peitgen, Heinz-Otto, et al. *Fractals for the Classroom.* New York: Springer-Verlag and NCTM, 1992.

Peterson, Ivars. "Escape into Chaos." *Science News* (May 26, 1984): pp. 328–329.

Peterson, Ivars. "Surreal Films: A Soapy Solution to the Math Puzzle of Turning a Sphere Inside Out." *Science News.* October 10, 1998. Volume 154, no. 15, pp. 232–234.

Posamentier, Alfred S., and William Wernick. *Advanced Geometric Constructions.* Palo Alto, Calif.: Dale Seymour Publications, 1973.

Rucker, Rudy. *The Fourth Dimension: A Guided Tour of the Higher Universes.* Boston: Houghton Mifflin Co., 1985.

Seymour, Dale. *Visual Patterns in Pascal's Triangle.* Palo Alto, Calif.: Dale Seymour Publications, 1986.

Smith, Sanderson M. *Agnesi to Zeno: Over 100 Vignettes from the History of Math.* Berkeley, Calif.: Key Curriculum Press, 1995.

Weeks, Jeffrey. *Exploring the Shape of Space.* Emeryville, Calif.: Key Curriculum Press, 2001.

Wills, Herbert III. *Leonardo's Dessert—No Pi.* Reston, Va.: NCTM Publications, 1985.

Zaslavsky, Lawrence. *Africa Counts: Number and Pattern in African Culture.* Brooklyn, N.Y.: Lawrence Hill Books, 1999.

Quilting References

Hargrave, Harriet. *Mastering Machine Appliqué.* Lafayette, Calif.: C & T Publishing, 1990.

Horton, Roberta. *An Amish Adventure,* 2d edition. Lafayette, Calif.: C & T Publishing, 1996.

Knox, Gerald M., ed. *Better Homes and Gardens American Patchwork and Quilting.* Des Moines, Iowa: Meredith Corporation, 1987.

Porter, Liz, and Mary Ann Fons. *Quilter's Complete Guide,* revised edition. Birmingham, Ala.: Oxmoor House, 2000.

Videos

"The Clifford Torus" is available through Media Magic at P.O. Box 508, Nicasio, CA 94946.

"The Hypercube: Projections and Slicing" by Dr. Thomas Banchoff is available from the author at a link through **http://www.keymath.com/quilts.**

"Life by the Numbers—A New Age." Pittsburgh: WQED, 1998.

"Life by the Numbers—Patterns of Nature." Pittsburgh: WQED, 1998.

"Life by the Numbers—Seeing Is Believing." Pittsburgh: WQED, 1998.

"Life by the Numbers—Shape of the World." Pittsburgh: WQED, 1998.

"The Shape of Space." Emeryville, Calif.: Key Curriculum Press, 2001.

Web Links

http://www.keymath.com/quilts

Isometric Dot Paper

Rectangular Dot Paper

Half-Centimeter Grid Paper

Sierpiński Curve Grids

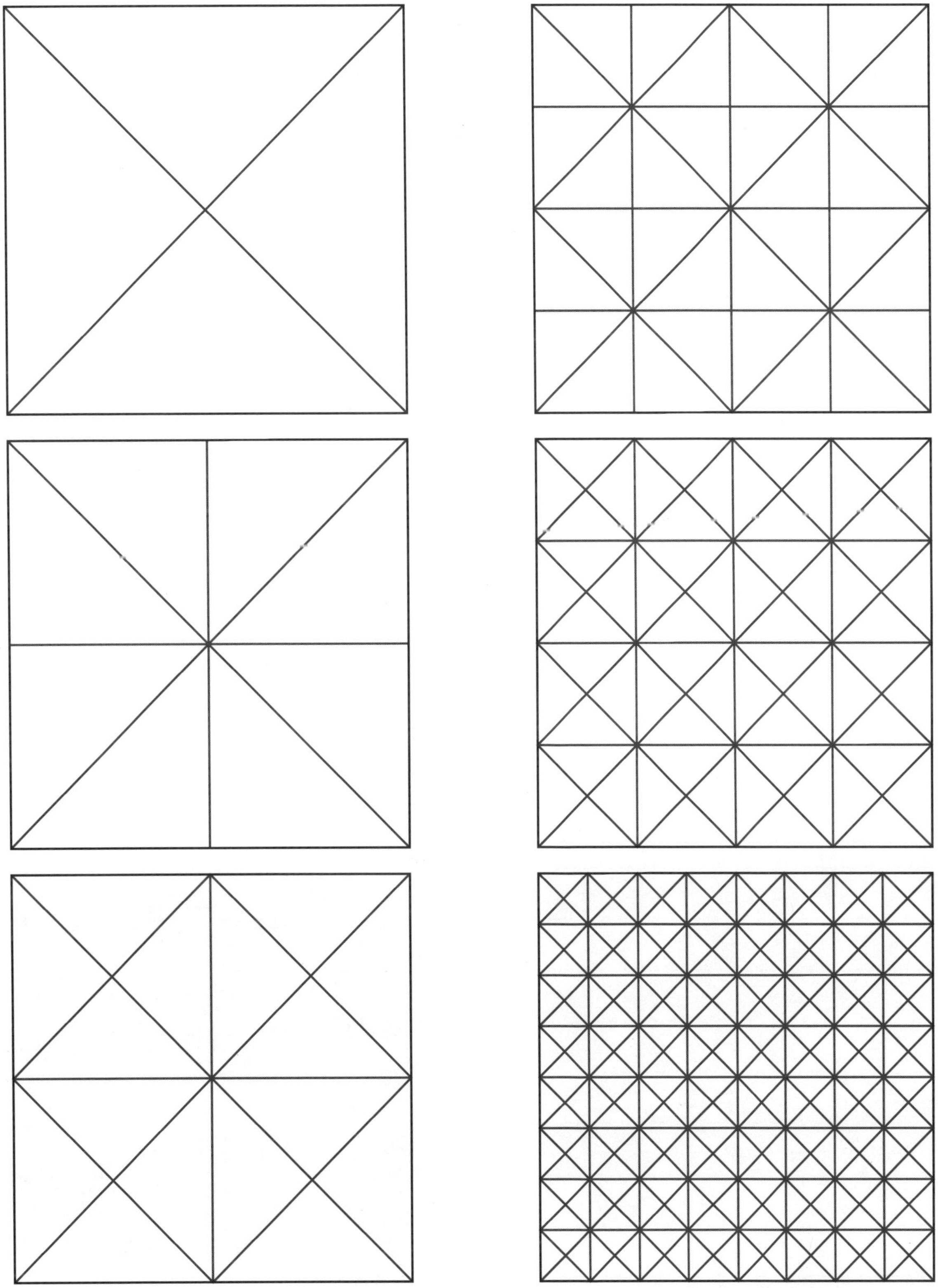

Hilbert Curve Grids

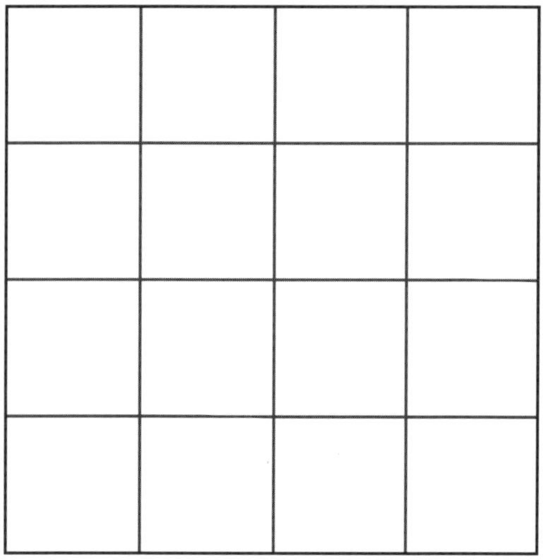

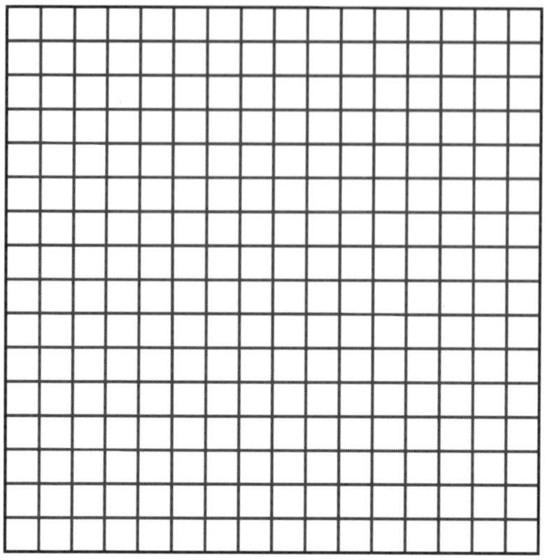

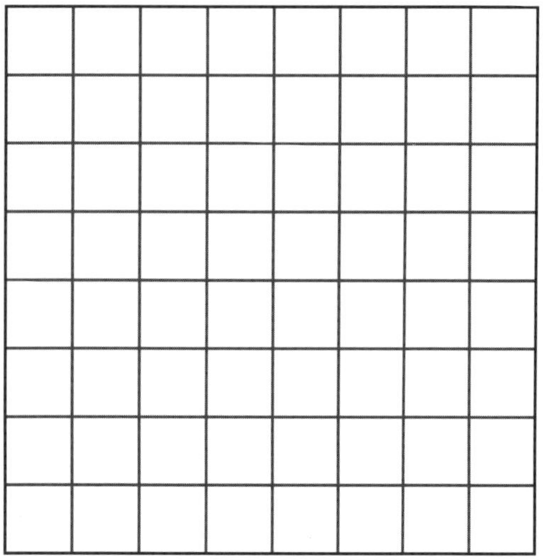

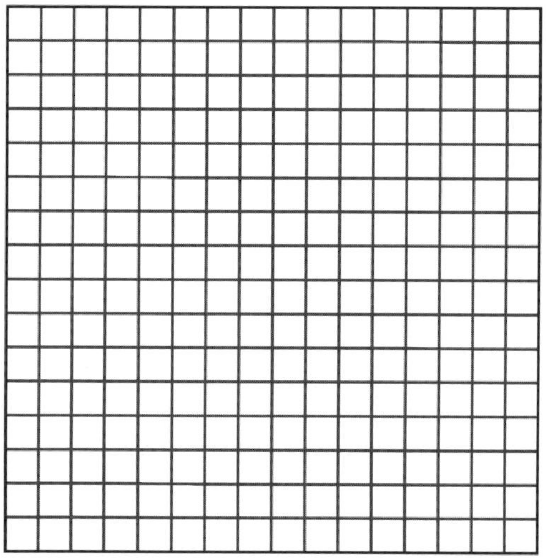

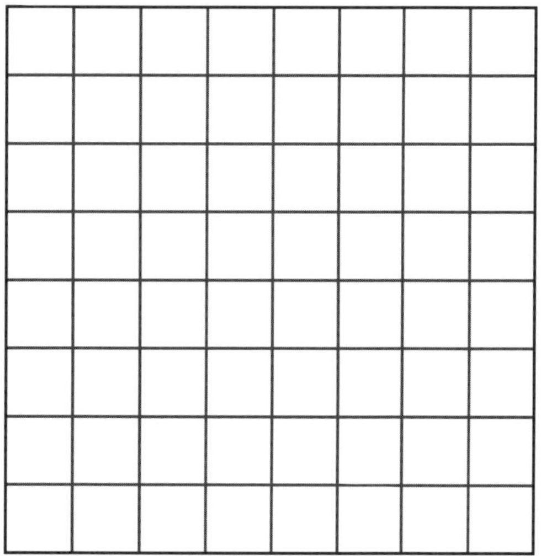

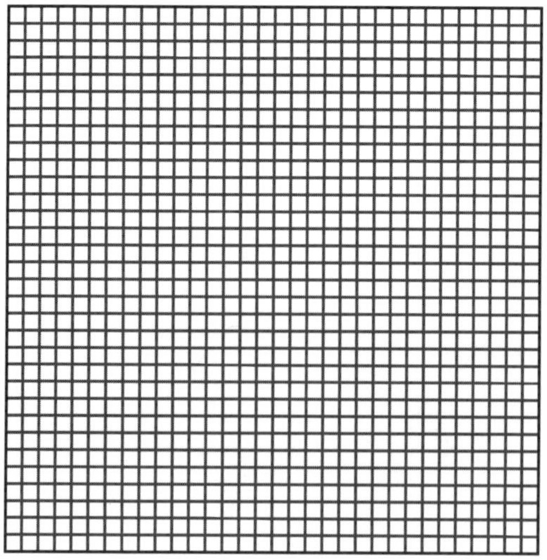

Tumbling Blocks Grid

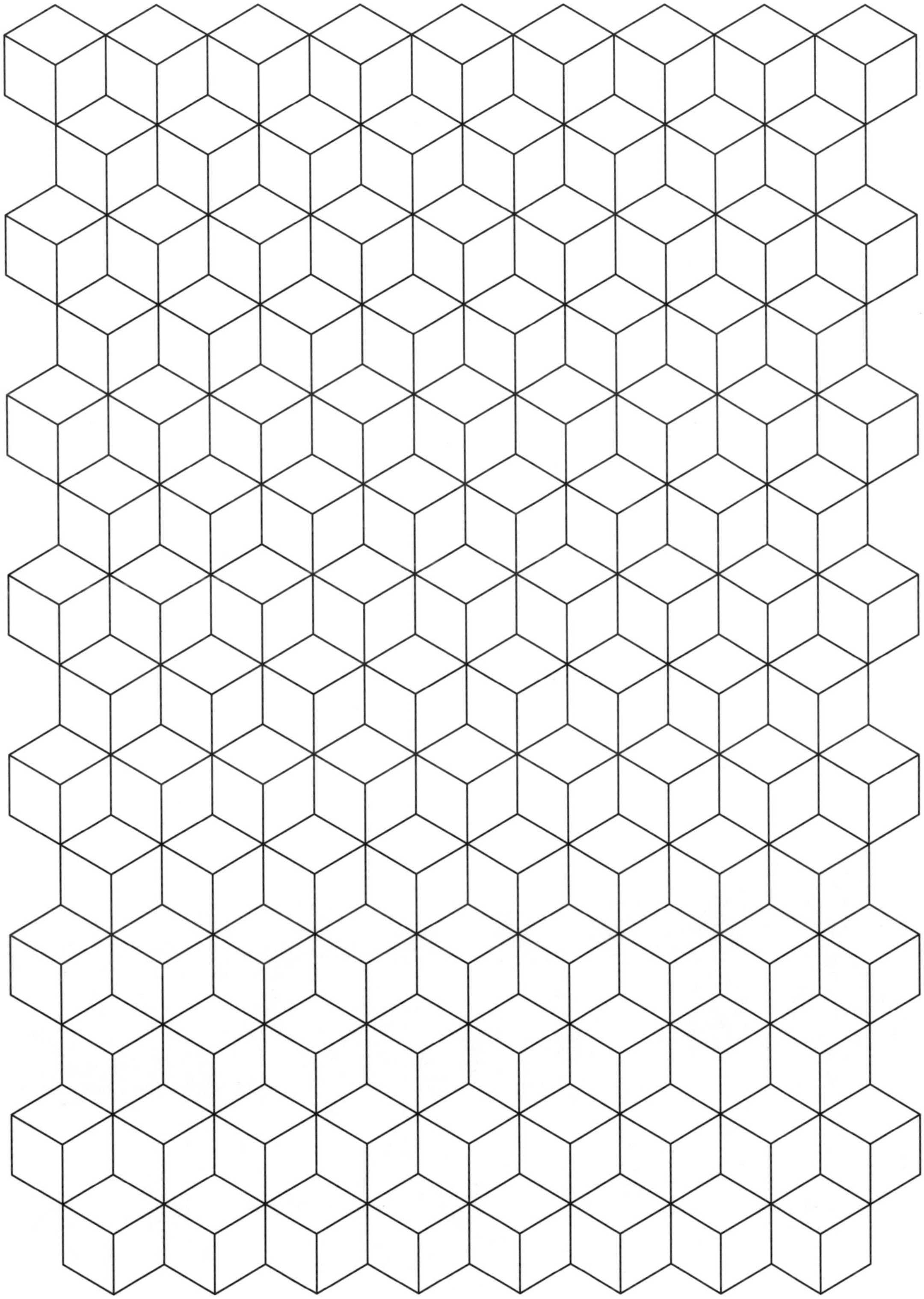